Simplicius
On Aristotle Physics 6

Simplicius on Aristotle's *Physics* 6

Translated by
David Konstan

Cornell University Press
Ithaca, New York

First published in 1989 by Cornell University Press

Library of Congress Cataloging-in-Publication Data

Simplicius, of Cilicia.
    Simplicius on Aristotle's Physics 6.

    (The Ancient commentators on Aristotle)
    1. Aristotle. Physics. Book 6. 2. Science,
Ancient. 3. Continuity—Early works to 1800.
4. Physics—Early works to 1800. I. Konstan, David.
II. Title.   III. Series.
Q151.A8S56   1988   530   88–47747
ISBN 0-8014-2238-8

63,286

Printed in Great Britain

Contents

Introduction
Richard Sorabji

Book Six of the *Physics* shows Aristotle at his best. Its subject is the continuum, and we can distinguish four main areas of discussion. First (chapters 1-4 and 10), he attacks atomism,[1] but an atomism of a rather extreme sort, according to which the atoms of which matter is composed, despite having a positive size, are *conceptually* impossible to divide. Democritus and Leucippus, who should be regarded as the first atomists, had not been conscious of the need to decide whether the impossibility of dividing their atoms was conceptual or physical. For some purposes, e.g. the solution of Zenonian paradoxes, they needed the impossibility to be conceptual, for others merely physical. But in the circle of Plato's Academy, at least in the person of Xenocrates,[2] atoms had been postulated which it would be conceptually impossible to divide. Aristotle does not ask himself the conceptual/physical question, but his attack is appropriate only against atoms of this conceptually indivisible sort.

One line of attack is to argue that there is no possible way to *arrange* such atoms, whether by contact, by continuity, or by succession, so as to build up the macro-world. This argument in 6.1, 231a21-b18, turns on the definitions of contact, continuity and succession. A second line of objection (6.1, 231b18) is that if extensions are composed of conceptually indivisible units, as some had supposed, then so are time and motion, which nobody wanted and which Aristotle argues to be impossible. Neither moving bodies (6.4, 234b10-20; 6.10, 240b8-241a6), as Plato had seen,[3] nor distances traversed (6.1, 231b18-232a17) can be conceptually indivisible. For when a body moves into an adjacent space, there must be a stage when *part* of it has entered *part* of the adjacent space, while *part* still occupies *part* of the original space. Both space and moving body must then have *parts*, and be conceptually

[1] The following discussion of atomism is based on Richard Sorabji, *Time, Creation and the Continuum (TCC)*, London 1983, chs 2 and 22-6.
[2] See the Platonist version of atomism which is expounded for attack in [pseudo-] Aristotle *On Indivisible Lines* ch.1, 968a1-b23, and the associated views of Xenocrates on indivisible lines, fragments 43-9 (Heinze), 128-47 (Isnardi).
[3] Plato *Parmenides* 138D-E7.

1

divisible. If they lacked parts, then a body could never *be* moving, but at least only *have* moved with a jerk, that is by disappearing from one indivisible space and reappearing in another. But such atomic motion would be paradoxical, in Aristotle's view, since, for one thing, the body would be *resting* in every single space (6.1, 232a12-17). And atomic time would be equally objectionable.

These are only two lines in Aristotle's attack, but they did much to shape subsequent atomism. We can see the Greek atomists Diodorus Cronus and Epicurus looking for a way of arranging atoms that escapes Aristotle's strictures.[4] We find them accepting atomic motion and atomic times as perfectly possible.[5] And Islamic atomists were later to agree.[6]

Since Aristotle believes that motion is not jerky, but continuous, he has to confront a second and related topic in *Physics* 6, that of first and last instants of change. An instant like twelve o'clock, or a 'now' as he would call it, differs from a time-atom in being sizeless, like a geometrical point. A representative problem, already found in Plato,[7] concerns stopping and starting. When something stops moving, what is the last instant of motion and what the first instant of rest? If these are the *same* instant, the body seems (absurdly) to be both in motion and at rest at the same time. If either instant precedes the other, there will (since instants cannot be immediately next to each other) be a gap during which the body is in both states or in neither. If we say that there is a last instant of motion, but not a first of rest, or vice versa, this appears arbitrary. What are we to do?

Aristotle's solution to this version of the problem (6.3, 234a34-b5) is that there are no instants of motion or of rest. An instant, being sizeless, cannot accommodate rest, or motion, or any kind of change (6.3, 234a24-b9; 6.6, 237a14-15; 6.8, 239a10-b4), for all of these presuppose a stretch of time. Admittedly, there are instants at which a body is *at* or *away from* the starting point or finishing point of a journey or of a change of colour, so we can raise a question about these. But the answer in this case (8.8, 263b11; 263b17-21) is that there are first and last instants of being *at* the terminus of a change, not first or last instants of being *away* from it (6.5, 235b7-8; 235b31-236a27; 6.6, 236b32-237b22; 8.8, 263b15-246a6).[8] This is not an arbitrary verdict, but flows from the nature of the continuum:

[4] Diodorus ap. Sextum *M* 10.120; Epicurus ap. Lucretium 1.605-6, discussed *TCC* 370-1; 373.
[5] Diodorus ap. Sextum *M* 10.85-92; 10.97-101; 10.119-20; 10.143; Epicurus ap. Sextum *M* 10.142-54; ap. Simplicium *in Phys* 934,26, discussed *TCC* 15-20; 347-8; 369-70; 375-6.
[6] Maimonides *Guide for the Perplexed* 1.73, discussed *TCC* 385; 394.
[7] Plato *Parmenides* 156C-157A.
[8] *TCC* 412-415. Aristotle seems to forget his principle that there is no first instant of being away, when he argues that a body must pause for a while before reversing

no point is immediately *next* to the terminal point of a change, or to any other point, for however small the distance that separates two points, there is always room for further points in between.

Aristotle's treatment of stopping and starting is important. Throughout Book Six of the *Physics* he presupposes motion at uniform speed. He does so, for example, in his argument that whether space is infinitely divisible or composed of indivisible units, time will have the same status (see 6.2, 232a28-31).[9] If he had once considered the possibility of continuous acceleration, he would have had to revise his arguments, and allow for the possibility that a body might have a different speed at every instant. But then he would have had to accept the idea of motion at an instant; by rejecting it, he is said to have 'bedevilled the course of dynamics'.[10] In fact, however, his treatment of instants at and away from a terminus is a positive help towards framing a good definition of motion at an instant, even though he could not foresee the utility that such a definition would have.[11]

A third topic in Book Six of the *Physics* is Zeno's four paradoxes of motion (6.2, 233a21-31; 6.9; 8.8, 263a4-b9). Bertrand Russell has said that they raise 'difficulties which it has taken two thousand years to answer, and which even now are fatal to the teachings of most philosophers'.[12] But even now I do not believe that there is an agreed solution, and I think that they can be solved only be accepting and learning some remarkable lessons about the continuum, which without Zeno one might never have observed.[13] The four paradoxes of motion are particularly appropriate to Aristotle's *Physics*, which is a treatise on nature (*phusis*), given that Aristotle defines a thing's nature as an inner cause of its *motion* (or other changes).

A fourth topic is the impossibility of infinite changes, other than

---

direction, on the grounds that there is not only a first instant of having *reached* the point of reversal, but also, unexpectedly, a first instant (which must be kept separate) of being *away* from it again (*Phys.* 8.8, 262a31-b3; b21-263a3: *TCC* 324). An analogous puzzle about the last point at which an approaching object is still invisible and the first at which it is visible is introduced by Aristotle at *Sens.* 7, 449a21-31, but turned against Aristotle by Diodorus Cronus, who argues that such puzzles show the need for atomism. For they can be solved if distances, or in the case of Diodorus variant sizes, never differ from each other by less than an atomic unit (Diodorus ap. Alexandrum *in Sens.* 122,21-3; 172,28-173,1: *TCC* 345-7; 416-17).

[9] Thomas S. Kuhn, 'A function for thought experiments', in *Mélanges Alexandre Koyré*, vol.2, Paris 1964, 307-34, repr. in his *Essential Tension*, Chicago 1977, ch.10.

[10] G.E.L. Owen, 'The Platonism of Aristotle', *Proceedings of the British Academy* 51, 1966, 125-50, at 148, repr. in his *Logic, Science and Dialectic*, London 1986, see p.218, and cf. 58-61; 160-1.

[11] *TCC* 414-16.

[12] Bertrand Russell, *Our Knowledge of the External World*, London 1926, 175.

[13] *TCC* ch.21.

the eternal rotation of the heavens (6.2, 233a28-b15; 6.7; 6.10, 241a26-b20).

Simplicius' commentary on the *Physics* is shown by cross-references to be the second of three commentaries by him on Aristotle which are extant and undisputed as to authorship, *in Cael.*, *in Phys.* and *in Cat.* (the *in DA* is disputed). They were written after the closure of the Athenian Neoplatonist school in AD 529, and after the departure of the Athenian philosophers in 532 from Ctesiphon, where they had briefly taken refuge with the Persian King Chosroes. It has previously seemed that Athenian Neoplatonism disappeared from the map of history, with Simplicius' commentaries as its final memorial, and it has been a matter of doubt and controversy where he can have written these commentaries, and whether he wrote them in solitude apart from any teaching activity. The commentaries are not systematically divided into lectures, like some of the Alexandrian ones, and they refer to readers rather than to an audience. But nevertheless evidence has recently been offered that Simplicius may have re-established Athenian Neoplatonism in Ḥarrān (Carrhae) in Upper Mesopotamia, in a school which lasted until the tenth century AD.[14]

Simplicius compares different manuscripts of Aristotle, and gives us the interpretations of major Aristotelians and Neoplatonists, including Aristotle's pupil Eudemus of Rhodes, whom he calls the best commentator (991,28-9), Aristotle's immediate successors Theophrastus and Strato, Aristotle's principal editor Andronicus, the Aristotelian Aspasius, the most famous of the Aristotelian commentators, Alexander, with whom, however, he repeatedly disagrees, the Neoplatonist Porphyry and the independently minded commentator Themistius, whom he particularly respects and calls clever (*euphradês*, 968,30). In this last passage, Simplicius sides with Themistius against Alexander in saying that Aristotle is wrong, and this fits with the account in his later *Categories* commentary of the duties of a Neoplatonist commentator on Aristotle. He should be impartial, neither misrepresenting what is well said, nor trying to exhibit Aristotle as infallible, as if enrolled in Aristotle's school (*hairesis*, *in Cat.* 7, 26-9), but for a more favourable attitude towards Aristotle, see *in Phys.* 1040,13. It has recently been suggested that all post-Aristotelian movements in Philosophy

---

[14] I. Hadot, 'La vie at l'oeuvre de Simplicius d'après des sources grecques et arabes', in her *Simplicius, Sa vie, son oeuvre, sa survie*, Berlin 1987, 3-39, incorporates the findings of M. Tardieu, 'Les calendriers en usage à Ḥarrān d'après les sources arabes et le commentaire de Simplicius à la physique d'Aristote', ibid., 40-57; 'Sabiens coraniques et sabiens de Ḥarrān', *Journal Asiatique* 274, 1986, 1-44; *Coutumes nautiques mésopotamiennes chez Simplicius*, in preparation.

recognised a founder or founders with whom, after the first generation, they did not disagree.[15] Simplicius considers this attitude normal within a school, but will not take it towards Aristotle, even if he would to Plato.

At 946,24-6, Simplicius modestly claims originality (cf. 1040,15-16) for presenting a puzzle and a solution concerning the sense in which time and magnitude can be infinite, given that Aristotle had denied (*Physics* 3, 5-8) that there could be a more than finite number of anything. Time is infinite, he suggests, only in the sense that a circle is. That is, it measures the movement of the heavens which, being circular, has neither beginning nor end. Consequently magnitudes too are infinite only in the sense that some of them are circular (946,15-16; 20-4). As a matter of fact, a similar idea about time had already been expressed by Proclus, who says:

> The advance of time is not, as it were, a line that is single, straight and infinite in both directions, but finite and circumscribed ...
>
> The whole of time is contained in a single revolution of the whole universe ...
>
> The revolution of the whole has as its measure the entire extent and development of time, and no extension is greater – *unless it be by repetition, for in that way time is infinite* [my italics].

Proclus is thinking of the great revolution in which the heavenly bodies eventually return to their original alignments. But the last twelve words reveal that in fact one can still think of time as a straight line of a more than finite length, so a better solution is needed.

Curiously, Simplicius omits to mention an idea about the infinity of time which he presents in his commentary on *Physics* Book Three (506,3-8). Time does not contain a more than finite number of years, because there is only one year that exists, namely, the present. Of course, this way of restricting the infinity of time would not suit Simplicius in the context of his Book Six discussion, because no parallel could be drawn with the case of magnitudes, which do not flow away.

Simplicius' commentary on Book Six is valuable for the information it gives us on Zeno's paradoxes, although some of that information is independently available from Themistius or Diogenes Laertius. One of the paradoxes denies that you can reach any destination, because you would first have to go halfway, then half

---

[15] David Sedley, 'Philosophical allegiance', in J. Barnes, M.T. Griffin (eds), *Philosophy and Roman Society*, Oxford 1988. But for some nuances, see L. Taràn, 'Amicus Plato sed magis amica veritas', *Antike und Abendland* 30, 1984, 93-124.

the remaining distance, then half the remaining distance, and so on
ad infinitum. On the version preferred by the commentators, you
would· *first* have to go halfway. Because the divisions are
half-distances, Aristotle calls the argument the dichotomy (*Phys.*
6.9, 239b22), and elsewhere (*Phys.* 1.3, 187a1-3), he tells us that
somebody tried to meet some dichotomy argument by positing
atomic lengths. In his comments on that earlier passage
(138,3-141,11),[16] Simplicius preserves the only direct quotations
from Zeno that we have. He refers to our dichotomy argument, but
distinguishes several others, and says that it was Plato's pupil
Xenocrates who was led by some dichotomy argument or other to
accept atomism. Reporting the interpretations of Alexander,
Porphyry and Themistius, Simplicius reveals that there was
controversy over whether the dichotomy in question was devised by
Zeno or (improbably) by Parmenides, or neither. In his comments on
Book Six, he fills out the paradoxes of motion and Aristotle's replies,
giving the interpretations of Eudemus and Alexander on the
stadium or moving rows (*Phys.* 6.9, 239b33-240a17), which is
particularly useful, because that is the hardest of the puzzles to
follow. He explains that the four may have been the only four, or the
most important four, on motion. He tells us the reaction of Diogenes
the Cynic to Zeno's denial of motion – getting up and walking. And
he supplies Achilles and the tortoise as the participants in the
puzzle which says that the fastest can never catch up the slowest, if
he gives him a start, because each time he reaches the slowest's
previous position, the slowest will still be ahead, albeit by a
diminishing fraction.

Although Aristotle rejects jerky motion, he distinguishes
elsewhere the case of light, heat and freezing, which can spread
jerkily, not part way before whole, but all in one go (*athroôs, Sens.* 6,
446a20-447a11; *Phys.* 1.3, 186a15; 8.3, 253b13-31). In ordinary
motion, as Theophrastus observes (ap. Simplicium 986,5-7), there is
a first instant of having reached the terminus, but not a first instant
of being away from the starting point. But he remarks, such special
cases as the spread of light behave quite differently. Here there *can*
be a first instant of light having spread (ap. Simplicium 107,12-16;
998,13-16; ap. Themistium *in Phys.* 197,4-9). Some post-Aristotelian
thinkers accordingly argued that jerky motion need not be the
preserve of the atomists. Once the Aristotelians had allowed jerky
progress for light, it could be postulated for ordinary motion, and
that not across atomic distances, but across distances of indefinitely
variable length. The advantage of postulating such jerky motion

[16] Much of this passage is translated by Jonathan Barnes, *Early Greek Philosophy*,
Harmondsworth 1987, 152-5.

was that it would solve Zeno's paradox of the half-distances (reported Sextus *M* 10.123-42; cf. *PH* 3.76-8; Damascius ap. Simplicium 796,32-797,2; *TCC* 53-4). Simplicius' teacher Damascius extends the idea of jerky progress to time itself (ap. Simplicium 796,34-797,13; Damascius *in Parmenidem* = *Dub. et Sol.* (Ruelle), 2.236,9-11; 2.237,21; 2.242,13-16, soon to be available in the edition of Westerink and Combès), although Simplicius demurs (797,27-36; *TCC* 52-61). There had been controversy, Simplicius tells us, even among those who did not want to introduce jerky motion: was Aristotle entitled to exclude jerky motion, given that he allowed the jerky spread of heat?

Aristotle might plead that some kind of continuity *underlay* the spread of heat. At any rate, he says in the *de Sensu* that although the discriminable shades of colour form a discontinuous series (6,445b21-9; 446a16-20), none the less colour shades and musical pitches have a kind of *derivative* continuity (445b26; b30; cf. *Phys.* 6.4, 235a18; a35-b6; 6.5, 236b5-10). He seems to have in mind that the *discontinuous* change to the next discriminable pitch or shade may be produced by the *continuous* movement of a stopper along a string, or by a *continuous* change in the proportions of earth, air, fire and water in a body. If it could be maintained in the same way that the discontinuous spread of heat was produced by some *continuous* underlying process, Aristotle might be able to claim that the spread of heat was after all no genuine exception. Here too, even if a whole region becomes hot at once, there would be no first or last instant of the underlying process of its becoming hot.

Aristotle draws attention to another type of discontinuous change at *Physics* 6.6, 237b10, involving indivisible entities like geometric points, and Simplicius records Alexander's explanation (997,30). When new end points are created by breaking apart a geometrical line, we can use the perfect tense and say that new points *have* been created, but there is no time at which the points *are being* created, so we cannot use the continuous tenses as we do when we say that something *is moving*, or *is changing colour*. At *Metaphysics* 3.5, 1002a28-b11, Aristotle extends this comment to the present instant, and thereby solves the paradox (*Physics* 4.10, 218a8-21) according to which the present instant can never cease, so that we are at the same present instant as the warriors of the Trojan war. The answer is that the present instant never is *ceasing*, but it can *have ceased*, and it will have ceased at any instant you like to take, however close after its existence. In this last respect, an instant's having ceased has something after all in common with having moved: there will be no first instant of its having ceased. Aristotle's recognition that the perfect tense is in some cases applicable, without the continuous present ever being so, was exploited against him by the atomist Diodorus Cronus. It should license the jerky motion which Aristotle rejected, for why should we not say that an atom *has* moved into an adjacent atomic space, without its ever being true to say, 'it *is* moving' (Sextus *M* 10,85-6; 91-2; 97-101).

Aristotle's insistence that motion and other changes are not jerky, but involve *part by part* progression, enables him to answer an extra puzzle, which he tacks onto the end of Zeno's paradoxes in *Physics* 6.9, 240a19-29. When a thing is changing from white to not-white, we need not agree that it is (absurdly) in *neither* state during the process. It will be *partly* in one state and *partly* in the other. But elsewhere in *de Sensu* 6, we have seen, Aristotle suggests a *different* way of avoiding jerky change by appeal to an *underlying* continuity. On the alternative, the change to non-white could still be viewed as a continuous process, even if the new shade did not spread part by part over the surface, but covered the whole surface at one go, and during the process, the surface would still be white.

Puzzles about the continuum were repeatedly used to question when something could possibly perish and even to argue for immortality.[17] Simplicius (983,27) reports a version from Alexander: when does Dion die, for it is neither when he is alive nor when he is dead? One answer would seem to be: at the instant of transition between life and death, but that only raises the question whether he would be alive or dead at that instant. Alexander thinks that Aristotle has a solution (983,30-984,2), and a suitably Aristotelian one would be that, if dying is thought of as a process, it has no first or last instant; there is merely a first instant of having died. If on the other hand death is thought of as instantaneous, there is no instant at which a man is dying, and no instant at which he did die, only instants at which he *has* died. It is the latter point that Alexander appears to make.

Simplicius' commentary tells us not only about the earlier history of Greek Philosophy, but also about his own period of late Neoplatonism, a period in which he thinks advances are being made (625,2; 795,33-5). One report concerned the jerky progress of time, but another concerns a digression introduced by Alexander into his lost commentary on the *Physics* (964,19-23). Some people accepted the Aristotelian view that the soul, or at least its irrational part, cannot exist disembodied, but differed from the Aristotelians in postulating special tenuous bodies as 'vehicles' for the soul. Alexander complains that the vehicle of our soul would have to be located in the same place as our flesh. It was a matter of great controversy in

---

[17] Diodorus Cronus ap. Sextum *M* 10.347-9; Sextus *M* 1.312; 9.269; 344-50; *PH* 3.110-14; Taurus ap. Aulum Gellium *Noctes Atticae* 7.13; Augustine *City of God* 13.11; Crescas *Intermediate Physics* 8.5.3; Moses Mendelssohn, as reported by Kant, *Critique of Pure Reason* ed. B, 413-14.

antiquity whether bodies could be in the same place:[18] it was a central tenet of Stoic metaphysics that they could, whereas Alexander upheld Aristotle's view that they could not. Coming after Alexander, Plotinus agreed with him that bodies cannot be in the same place.[19] But subsequent Neoplatonists wanted to allow exceptions, and Simplicius tells us that the celestial spheres penetrate all the way down to the centre of the earth, coinciding with each other, and with the sublunary elements on the way (531,3-9; 616,23-617,2; 623,32-624,2; 643,18-26; 966,11-12). On the other hand, he doubts if the soul's vehicle has to be in the same place as our flesh, or whether our soul even needs an additional vehicle, so long as it has the flesh to reside in (966,5-8).

Aristotle's *Physics* is a book that could reasonably be chosen by a philosopher who was allowed only one book on a desert island. Simplicius' commentary, with its 1366 pages in Diels' edition, is the largest of all the ancient commentaries on Aristotle. If it incorporated Aristotle's *Physics* and added its own panoramic view of the history of the subject throughout antiquity, it would be a still better choice. In fact, Simplicius tends to presuppose Aristotle's text, and does not expound it before commenting. But this disadvantage is overcome in the present translation by the summaries of Aristotle's argument which David Konstan has supplied, and this practice will be followed in translating other books of the same commentary.

Acknowledgments

The present translations have been made possible by generous and imaginative funding from the following sources: the National Endowment for the Humanities, Division of Research Programs, an independent federal agency of the USA; the Leverhulme Trust; the British Academy; the Jowett Copyright Trustees; the Royal Society (UK); Centro Internazionale A. Beltrame di Storia dello Spazio e del Tempo (Padua); Mario Mignucci; Liverpool University. I further wish to thank Eric Lewis and John Ellis for assistance and advice on this volume, I. Hadot for access to her bibliography on Simplicius and Titos Christodoulou for his bibliographical work.

---

[18] Richard Sorabji, *Matter, Space and Motion*, London and Ithaca N.Y. 1988, chs 5-7.
[19] Plotinus 2.7.2; 4.7.8[2].

Translator's note

The translation offered here aspires to literal accuracy, employing the same English word for all occurrences of a given Greek term wherever possible, and indicating by the use of square brackets all supplements that fill out the sense of the original. In most cases, these are wholly uncontroversial, but it seemed best that the reader be able to identify each instance. Summaries of Aristotle's argument have also been added in square brackets, to supplement the fragmentary lemmata provided by Simplicius.

John Ellis, David Furley, Charles Hagen, Eric Lewis, Alain Segonds, David Sedley, Cecilia Trifogli, J.M.O. Urmson and L.G. Westerink read all or part of the translation in manuscript, and I am profoundly grateful to them for saving me from numerous infelicities and errors. I wish to thank Gavin Betts, Leon Blackman, Pierre Habel, Sharilyn Nakata, and Walter Stevenson for assistance in compiling the index. To Richard Sorabji I owe a special debt for his constant help and encouragement. It is a pleasure to offer these colleagues my sincerest thanks.

Simplicius

On Aristotle Physics 6

Translation

The Commentary of Simplicius the Philosopher: Part Six, on Book Zeta of Aristotle's *Physics*

Since the Peripatetics have the custom of labelling the order of their   923,1
books according to the order of the letters of the alphabet, [that is,]
alpha, beta, gamma, and so on, they reasonably label the sixth   5
[book] of the *Course on Physics* zeta, which in [the case of] numbers,
indeed, signifies the number seven, but in [the case of] letters
occupies the sixth position.[1] It was remarked earlier [cf. 4,14;
801,13] that they call the five books before this one the *Physics*, and
the next three *On Motion*; for so Andronicus stipulates in the third
[book] of *The Books of Aristotle*,[2] and Theophrastus too bears   10
witness concerning the first group: when Eudemus [fr. 6 Wehrli]
wrote to him concerning one of the faulty manuscripts in respect to
the fifth book, he said: 'As for what you asked me in your letter to
write out and send [you] from the *Physics*, either I misunderstand or
the middle of the [passage], "Of the things that are motionless, this
alone I call 'being at rest'; for rest is the contrary of motion, so that it
would be a privation in that which can have it" [226b14-16],   15
contains something quite small.'[3] Thus, Theophrastus considered
the fifth book too to be 'from the *Physics*'. Aristotle himself says in
the eighth [book], toward the beginning: 'Let us begin first with
what we have defined earlier in the *Physics*. We say, then, that
motion is the actuality of the movable qua movable' [251a8]; he said   924,1
this in the third [book (202a7)]. And again: 'For nature was posited
in the *Physics* as a principle of motion and of rest' [253b7]; this he
said in the second [book (192b20)]. At the end of the eighth book he
says: 'That it is not possible for there to be an infinite magnitude has
been proved earlier in the *Physics*' [267b20]; he talked about this too   5

---

[1] Ordinarily, the archaic sixth letter of the alphabet, digamma (F) was used to designate the number 6, and zeta (Z) designated the number 7.

[2] Andronicus of Rhodes (first century B.C.) was responsible for arranging the books of Aristotle in the form in which they have come down to us. See I. Düring, *Aristotle in the Bibliographical Tradition*, Göteborg 1957, 413-25, and, on the title of Andronicus' book, pp. 421-2; Düring supposes that the titles according to Simplicius' text will have been *(Peri) Aristotelous bibliôn*, but it seems to be cited simply as *Hoi Aristotelous biblioi*.

[3] This passage is difficult; Diels interprets: '...or it differs very little from [the vulgate, i.e.] "Of the things...which can have it".'

in the third [book (204a8ff.)]. That they used to call the [first] five
[books] the *Physics* is evident from these things.

That they called the last three *On Motion*, Aristotle clearly
indicates in the first [book] of *On the Heavens* [272a28ff.] when he
says: 'But that is obvious, that it is impossible to traverse an infinite
[line] in a finite time. In an infinite time, consequently; for this was
10 proved earlier in the [books] *On Motion*.' And again [275b21]: 'the
argument in the [books] *On Motion*, that nothing that is finite
has infinite power and nothing that is infinite finite [power].' These
matters were discussed in the three [last books]. That the three
[books] are those *On Motion*, and the five [books] the *Physics*, Damas,
who wrote *The Life of Eudemus* [fr. 1 Wehrli],[4] also testifies, when he
says: 'and of the three [books] *On Motion* from Aristotle's treatise *On*
15 *Nature*.'[5] Now, they used to call *Physics* not only the eight [books], but
also the [books] *On the Heavens* and *On the Soul* and many others, but
specifically [just] the five [books] of the *Course on Physics*.

That the book that is now before us comes after the fifth in order,
Eudemus [fr. 98 Wehrli] again indicates when, closely following
Aristotle's [words], he attaches to what was said in the fifth book the
[statement]: 'Nothing that is continuous is composed of partless
20 things' [cf. 231a24, 233b31]. Andronicus too assigns this order to
these books, and indeed Aristotle, right at the beginning of this book,
uses [terms] that had been previously explained in the fifth book,
[namely] 'continuous', 'touching', and 'consecutive'. That this book
also comes before the seventh is obvious from the things which he
proves at the end of this book, [namely], that nothing partless moves,
25 and that no change is infinite [cf. 240b8ff., 237b23ff.], [since] he uses
these things in [Book] Eta, which indeed is seventh and next [after]
this one, as [already] proved. Similarly, that there is no infinite
925,1 motion in a finite time, whether the moving thing is infinite or finite,
he also proves here [in Book Six], but uses in the latter book as
proved.

231a21-29: 'If the continuous [*sunekhes*] and what is touching
[*haptomenon*]...' to 'for the extreme [*eskhaton*] and that of
which it is the extreme are different.'

[Nothing continuous can be made of indivisibles, since
indivisibles can neither be (i) continuous with each other (i.e.
with their extremes united) nor (ii) in contact (i.e. with their

---

[4] Nothing further is known of Damas (cf. Wehrli, p. 77).

[5] Reading *triôn* with Wehrli in place of the MSS' *tria*, which would give 'the three
books of those on motion'; but there are only three in all. Diels recommends (in the
apparatus criticus) emending *tôn* to *ta*, which gives 'the three books'.

extremes touching); for partless things (= indivisibles) have no
extremes.]

Given that there are five magnitudes in things, [namely], line,   5
surface, body [i.e. volume], and, in addition, motion and time, that
each is composed of those parts into which it is also divided has
seemed to everyone a common notion [i.e. an axiomatic or
universally shared idea]. Individually, however, some believe that
any magnitude that is taken is divided to infinity into [further]
magnitudes, on the grounds that the division never ends up in
partless things [*amerê*] – and on this account they also say that   10
magnitudes are composed of parts but not of partless things; while
others, who had given up on [the idea of] cutting to infinity on the
grounds that we cannot [in fact] cut to infinity and thereby confirm
the endlessness of the cutting, used to say that bodies consist of
indivisibles and are divided into indivisibles. Leucippus and
Democritus [68 A 13 Diels-Kranz], however, believed not only in   15
impassivity [or imperviousness (*apatheia*)] as the reason why
primary bodies are not divided, but also in smallness and
partlessness, while Epicurus [fr. 268 Usener] later did not hold that
they were partless, but said that they were atomic [i.e. uncuttable
(*atoma*)] by virtue of impassivity [alone]. Aristotle refuted the view
of Leucippus and Democritus in many places, and it is because of
those refutations in objection to partlessness, no doubt, that
Epicurus, coming afterwards but sympathetic to the view of   20
Leucippus and Democritus concerning primary bodies, kept them
impassive but took away their partlessness, since it was on this
account that they were challenged by Aristotle. Here, however,
Aristotle proposes to prove in general that magnitudes having a
continuous extent [*diastasis*] have not been constructed out of
partless things. First he demonstrates this in the case of the
magnitudes of a body, and then for motion and time. Since there are   25
also certain magnitudes that are composed of some touching [parts],
such as a house or a heap, he demonstrates simultaneously that
neither can such a magnitude arise out of partless things. If every
magnitude, accordingly, is composed either of continuous or of
touching parts, but partless things can neither be made continuous
with one another nor touching, [then] no magnitude would be
composed of partless things. And he proves first for the line and for 926,1
points that it is impossible for the line, which is continuous, to be
made of points. This is how he proves it: The line, being continuous,
is made up of continuous parts; those things were [said to be]
continuous whose extremes [*eskhata*] are one [i.e. in common]; if,
accordingly, the line is made of parts that have extremes [that are]
one, but points do not have extremes [that are] one, then lines can   5

not be made of points, according to the second figure of inference.[6]
Now, it is obvious that the line is continuous and made up of
continuous parts, the extremes of which are one. That the extremes
of points are not one, he proves from the fact that, for all things for
which there is an extreme, the extreme is one thing, and the thing of
10   which it is an extreme another. If, accordingly, the point is partless
and there does not exist any extreme of it, the extremes of two points
can not be one, such that some one thing is made out of them.
Consequently, it is impossible for any continuous thing to be made
out of points.

   If someone should say that a limit [*peras*] is not a part, and thus
nothing prevents there being a limit to a partless thing, even he
[must] agree that the whole and the limit are not the same thing.
15   Accordingly, there is something else again alongside the limit; but
in the point there is not. In general, what has a limit also has a
beginning and a middle, and so a portion [*morion*] as well.

   It has been proved, accordingly, that it is impossible for what is
continuous to be made of partless things. For what has been said
about the line will be said as well about other continuous things; for
it is not because they are points, but rather because they are
partless, that what has been said about them is true. Also partless
20   are nows [*ta nun*, i.e. temporal limits or instants] and units, so that
even if some people should say that there are partless bodies, as
Diodorus [Cronus] thinks he proves, the same will be said about
them as well.[7]

   231a29-b6: 'Furthermore, it is necessary either that points be
   continuous' to 'also separated in place.'

   [Since indivisibles are partless, they cannot touch each other
   part to part or part to whole, but only whole to whole; but such
   a fusion again yields an indivisible rather than a continuous
   thing, which by definition has distinct parts and is divisible.]

   Since it seems that certain continuous magnitudes are made of
25   [parts that are] not continuous but touching, like a wall out of
bricks, [Aristotle] proves that not even in this way can a magnitude
be made of points or of partless things at all, assuming in advance

   [6] The second (or indemonstrable, or hypothetical) figure of inference, in the Stoic
classification of propositional inferences, is in the form; if A then B; but not B;
therefore not A. Simplicius refers to this figure also under the name of potential
syllogism, or syllogism in the second figure. It may be noted that Simplicius has a
penchant for recasting Aristotelian arguments in this form.
   [7] Diodorus Cronus (*c.* 300 B.C.) was associated with the Megarian school of
philosophy, and a master of dialectical argument.

that, if something continous is made of points, it is necessary that it
be out of such as are in continuity or touching, and that they are not
continuous by virtue of what has been proved before. He then adds
that not even when they are touching do points make a continuous
thing. It was easy to prove that points also do not touch each other,   30
since those things touch whose limits are together [*hama*], but
points, which are limits, do not have a limit. But having laid out by   927,1
[logical] division the ways in which things may be said to touch each
other, and having said that either whole touches whole as point
[touches] point or as do lines superimposed [*epharmozousai*] on one
another; or one thing by a portion and the other as a whole, the way
a line [touches] a point; or [both] by portions as a line [touches] a line   5
at a point (for a point is, somehow, a portion of the whole), he adds
that, since points are partless, they will touch upon each other
neither by parts nor as whole to part, but rather, if [at all, then] as
whole to whole, so as to be superimposed upon one another. But
these could not strictly be said to be touching, since things that
touch were those whose limits are superimposed upon one another,
rather than [things that are superimposed] whole upon whole.
    Even if one were to consent, moreover, to saying that these things   10
too are touching or are adjacent to [*pelazein*] each other in some
other way, no magnitude would be made out of such things. For a
magnitude that is made of continuous things as well as one that is
made of things that touch has parts, 'one here and another there',
but that which is made of partless things has none. Thus, no
magnitude would be made even out of points that are touching. The
first proof proceeded from points not having limits, this one from   15
their not having parts. Limit is one thing and part another, even if
[Aristotle] himself seems here to have treated the limit as a part.
Having said that 'the continuous has one part here and another
there', he adds: 'and it is divided into parts that are different in this
way, that is, spatially separated.' For the continuous, since it is
neither composed of partless things nor divided into them, is divided   20
into continuous things. Thus, not only the whole, but also each of the
parts that have been divided out of it has one part here and another
there, separated not only rationally, like the qualities of an apple,
but also spatially.

231b6-10: 'But neither will point be consecutive upon [*ephexês*]
point' to 'and [between] nows time.'   25

[Neither (iii) can consecutive indivisibles produce a continuous
thing; for consecutive things are separated (if at all) by
something different in kind, but what is between indivisibles

in a continuum is the same in kind as they.]

Having demonstrated that no single magnitude can arise either out
of continuous points or nows[8] or out of ones that touch, he proves for
good measure that if anyone believes that any length or time is
928,1   made of consecutive points or nows, he is mistaken: first of all
because, even if these things were consecutive to one another, they
would not make a single thing, since there would be something of a
different kind in between, and second because it is not only
impossible for points to be continuous with each other or touching
one another, but it is not even possible for them to be consecutive to
5   one another. (The argument is also a fortiori. For if it is impossible
for points to be consecutive to one another, it is all the more
impossible that they touch one another, since things that touch, in
addition to being consecutive, also have their limits together. And it
is still more impossible that they be continuous, if it is necessary
that their limits run together into one with their parts each in
another place, standing apart in space.) That points are not
10   consecutive to one another, he proves by making use of the
definition of consecutiveness, once again employing the second
figure of the syllogism as follows: those things are consecutive
'between which there is nothing that is akin', as houses are
consecutive with one another between which there is not some
[other] house, even though a tree or something else might happen to
be in between, or an empty space. But between points there is
something of the same kind, [namely,] 'a line, and [between] nows,
15   time'. It is inferred, accordingly, that neither points nor nows are
consecutive. Now, it is obvious that between two points there is
everywhere a line and between two nows, time. But how is the line
the same kind of thing as points? [We may reply,] as the limit of that
of which it is the limit, just as he proved in the *Ethics* [1175a22ff.]
that the proper pleasure is akin to the activity of which it is the
limit. But if the point is the same kind of thing as the line, and the
20   now as time, and [if] the line and time are quantities, then the point
and the now would also be quantities, unless, after all, these too
come under [the category of] quantity on the grounds that they are
limits of a quantity. For under what[9] other category would they
come, if indeed everything in existence must be referred to one of the
ten categories? Now, if in every [stretch of] time there is a now and

[8] Aristotle uses the abverb 'now' together with the definite article in a technical
sense to mean a point or instant of time; because the argument sometimes depends
on the connection between this technical sense and the ordinary, looser sense of 'now'
as 'the present time', I have chosen to render Aristotle's term throughout as 'now' or 'a
now'.
[9] Reading the interrogative adjective *tína* instead of the indefinite *tina* (Diels,
MSS), i.e. 'they would fall under some other category'.

in every line a point, then any things that have line or time between
them would have a point or now between them, unless indeed the   25
point is in the line, and the now in time, potentially and not actually;
but what exists potentially can also come to be in actuality.

Eudemus [fr. 99 Wehrli] treated the argument in the following
way: 'For if,' he said, 'partless things are consecutive, then it is
absolutely necessary that something not of the same kind be between
them, so that there would not be a point, but [rather] a line or empty   30
[space] between points in a length. [If a line[10]], accordingly, the line
will not be made of the points; for the consecutive points are not in it.
But if there is empty [space], there will be more empty [space] in the   929,1
continuous [stretches] than the things of which they are made, that
is, the so-called consecutive points; or [rather], there will be no
magnitude at all. For just as two points that are touching produce no
length, so neither do a point and empty [space].' [We may say that]
Eudemus, accordingly, correctly understood what was said by
Aristotle, [namely] that he was not denying absolutely that point is
consecutive to point or now to now, but rather that they were   5
consecutive in such a way that a length or [stretch of] time was made
of them. For if between the points or nows there is line or time, how
could the line be composed of the consecutive points, if indeed the
consecutive points are only its limits?

Alexander argued the point in the following way as well: 'It was in   10
accordance with their own [view (i.e. that of Leucippus and Democritus)],' he said, 'that [Aristotle] stated that the line was the same kind
of thing as points and time as nows, if, that is, each of them is
composed of partless [things], as they themselves say, the line of
points, and time of nows. So that it is according to them themselves
that points are not consecutive, if, that is, what is made out of them is
(according to them) between them, and is of the same kind as they   15
are.' Alexander seems to have believed that Aristotle denied that they
are consecutive in every respect, and not [merely] that they are
consecutive in such a way that length and time are made up of them.
But this indeed is what Aristotle clearly meant by his own words.

231b110-15: 'Further, it would be divided into indivisibles' to
'or into indivisibles. And this is the continuous.'   20

[If a continuous thing were made of indivisibles, it would be
divisible into indivisibles; but no continuous thing is divisible
into partless things. Further, there can be nothing different in
kind between indivisibles, for such a thing must itself be either

---

[10] Supplement adopted from Aldine (noted by Diels in the apparatus criticus).

reducible to indivisibles or forever divisible, and the latter is
continuous.]

I believe that this argument too is in proof of the same thing, [that
is], that partless things are not even consecutive in such a way that
length or time is made up of them. The [word] 'further', which is
added, also indicates that this argument too is offered in regard to
the same thing. [The argument] goes like this: if the continuous,
25   that is the line and time (which is what [the word] 'each' [in
Aristotle's text] refers to), is made of consecutive partless things, the
one out of points, the other[11] out of nows, then 'it would be divided',
too, into consecutive 'indivisibles'; for each thing is also divided into
those things out of which it is composed. That is the demonstration
of the conditional premise. The additional premise, next, is by way
of denial of the consequent: 'but it was [said] that no continuous
30   things were divisible into partless things.' The conclusion or result,
since it is clear, is omitted, [namely] that the continuous is not
composed of consecutive partless things. He proves the additional
premise that states that no continuous things are divisible into
partless things by substituting [the term] 'between' for 'continuous'
(for what is between was the line and time [i.e. continuous things]),
930,1   and he says that 'there cannot be another kind [of thing] between'
the indivisibles, other than the divisible or the indivisible (for
between these there cannot be any grey or lukewarm), especially if
5   what is between is said to be composed of indivisibles. For [logical]
division in this kind [of thing, that is], of the divisible and
indivisible, is into contradictory pairs. If, accordingly, the
continuous [stretch] between is not indivisible – since it is said to be
composed of partless things, and that which is composed is also [able
to be] divided – it is necessary that it be divisible, and that either
'into indivisibles or into forever divisibles'. But for its not being into
indivisibles he next adds the reminder: 'for if it is into indivisibles,
10   there will be indivisible touching indivisible', so that some one thing
is made out of them. But this indeed was demonstrated to be
impossible.[12]
        Alexander inquires how the proof [can be] sound, if on the one
hand [Aristotle] proves that, because the continuous is not divided
into partless things, neither is it composed of partless things, when
he says: 'but it was [said] that no continuous thing is divisible into
partless things', while on the other hand he proves that it is not
15   divided into partless things on the grounds that partless things do

[11] Reading *ho* instead of *hê* (as suggested by Cecilia Trifogli).
[12] Simplicius' interpretation of the last clause in Aristotle's text is forced, in the
effort to support his thesis that the whole passage concerns consecutive partless
entities.

not touch one another, that is, are not compounded with one
another, by virtue of which he already proved that a magnitude
cannot be composed of partless things. He resolves the objection by
stating that the argument from what is between is different – the
one that says that [something] is either divisible or indivisible, 'and
if divisible, either into indivisibles or into forever divisibles'. If,
accordingly, the continuous is neither indivisible, nor divisible into   20
indivisibles, then it would of necessity be divisible into forever
divisibles. He says: 'This is confirmed from the following. If it should
be divided into indivisibles, "there will be indivisible touching
indivisible", such that a continuous thing is made.' [We may say
that] neither is the proof circular as it was stated previously, but
rather [Aristotle] first proved – on the grounds that partless things
do not touch each other – that continuous things are not composed of
partless things (not that touching and being composed or rather   25
being in continuity are the same thing, but that touching is
altogether prior to being in continuity), while here he proves, on the
same grounds again, that continuous things are not divided into
partless things, and from this [in turn] that they are also not
composed of partless things. Thus, both are proved on grounds of not
touching, but [the fact of] not being composed [is proved] also on the
grounds of not being divided, and this proof is not circular. And [we
may say that] in such cases neither is a circular proof absurd. For if   30
it is not divided into partless things, it is not composed of partless
things, and if it is not composed of partless things, it is not divided
into partless things. And so for all things that are mutually implied
by one another, for if she has given birth, she has milk, and if she
has milk, she has given birth.

Alexander adds also the following kinds of reasoning which he
took from Eudemus [fr. 100 Wehrli]: 'If a magnitude is made of   35
partless things, there will be a line greater than [another] line by a   931,1
point. But if this [is the case], either not every line will be [able to be]
divided in half, or if [every line] should be [so] divided, then a [line]
that is composed of an odd number [of points] would also be divided.
Thus it would happen that the point too is divided. Furthermore, if
in general a line were bigger than [another] line by a point, a circle
would also be [bigger] than [another] circle [by a point], so that one
would be made of an even number, the other of an odd number [of
points]. Accordingly, either we shall not [be able to] divide the one   5
made of an odd number [of points] into semi-circles, or we shall
divide a point in half.'[13]

[Aristotle] set up the [logical] division of the divisible into [either]

---

[13] Cf. Themistius 182,24- 183,1, from which it appears that Eudemus is referring to
the area, not the the circumference, of circles and semi-circles.

indivisibles or forever divisibles, so that he might comprise the continuous in that which is divisible into forever divisibles.

> **231b15-18** 'It is evident that every continuous thing is divisible' to 'the extreme of continuous things is one and touches.'
>
> [Every continuous thing is forever divisible; for if it were divided into indivisibles, then indivisibles must touch each other.]

10 Having proved in the earlier arguments that the continuous is not composed of partless things, he now proves that neither is it divided into partless things. On the one hand, it is obvious that a continuous thing that has magnitude is divisible, and that it is necessary that what is divided be divided either into divisibles or into indivisibles, as was said before. For there is not anything between the contradictory pair. On the other hand, that the continuous is not

15 divided into indivisibles, he proves on the grounds that this would result in indivisible touching indivisible. For if the extremes of those things that make up the continuous become one, it is necessary first that they touch one another, and that they thereby coalesce into one. Thus, if they become one, they also invariably touch; if, however, they touch, they do not invariably [become] one. For this reason he

20 spoke of becoming one as a final cause [cf. 194b32-95a3] of touching: for magnitudes that do not touch cannot become one. That it is impossible for something continuous, that is, a magnitude, to be made of partless things that touch, he proved earlier by the [logical] division of things that touch. For neither is it possible for partless things to touch by a part, nor for parts touching whole to whole and superimposed upon one another to produce something continuous,

25 because the continuous has different parts in different places. Alexander says, 'a magnitude composed of touching things is more commonly called continuous', and thus he understands Aristotle as saying 'for the extremes are one, inasmuch as the extremes of continuous things touch'. But [we may reply that] Aristotle did not call something made of touching things continuous, but said rather that the [common] extreme of continuous things that make a greater continuous thing out of themselves is one. If it is one, then it is also

30 necessary [for the extremes] to touch, as was said. For things that have not first touched do not become one.

932,1   **231b18-232a17**: 'By the same reasoning magnitude and motion and time' to 'if they are not motions, [it would be possible that] motion not be made of motions.'

[Either magnitude, time, and motion are all alike made of indivisibles, or none is. Let each indivisible magnitude A,B, C..., correspond to an indivisible motion D, E, F.... Something that has moved somewhere was moving there earlier, but over a partless magnitude what has moved cannot previously have been moving, for it cannot have vacated part of one partless magnitude and entered part of another. So it must proceed, if at all, by indivisible 'moves', that is by disappearing from one partless magnitude, and reappearing without delay in another. Further, if a thing is not moving at each indivisible magnitude, it must be at rest at each; again, a thing will be at rest when motion is present, if the indivisible parts of a motion are motions.]

Having proved in general that nothing continuous is composed of indivisibles or divided into indivisibles, since there are, to put it briefly, three [kinds] of continuous things – magnitude, motion and   5
time – he now proves that the same reasoning applies to them. Either all of them alike are composed of indivisibles and are divided into indivisibles or none of them is. First he proves that motion is similar to magnitude, and then that time too takes the same reasoning as these. That if magnitude is composed of indivisibles, then a motion over that [magnitude] will also be constructed out of   10
[just] as many indivisible motions, he proves by first clarifying the problem, as is his custom, by an illustration [*ekthesis*].[14] For if there were a magnitude ABC composed of three indivisible parts ABC, and if there were a thing O moving over the magnitude ABC [with] a motion DEF, then this [motion] too will be composed of certain indivisibles. That is the illustration of the problem which exhibits   15
the similarity. The demonstration is according to the second hypothetical [figure of inference (cf. n. 6)] by denial of the consequent. The syllogism is as follows: if the magnitude is made of indivisibles, then the motion is also made of indivisibles; but in fact the motion is not made of indivisibles, as I shall prove; neither, consequently, is the magnitude. He has posited the conditional premise out of what has been proved earlier. For if the motion   20
conforms to the magnitude over which it occurs, as has been proved, then if the magnitude were composed of indivisibles, the motion over it would logically [be so] as well. He establishes the additional

[14] Simplicius commonly uses the term *ekthesis* to refer, as here, to a kind of diagram, but he apparently understands the diagram, in turn, to represent a particular or illustrative example.

premise that says 'but in fact the motion is not made of indivisibles'
by another syllogism as follows: if motion is made of indivisibles,
then moving [*kineisthai*] is also made of indivisibles; but this, in
25   fact, is impossible; and so, consequently, is the antecedent. Again,
the conditional premise that says 'if motion is made of indivisibles,
then moving is also made of indivisibles' he proves by syllogism
according to the second indemonstrable [figure of inference (cf. n. 6)]
thus: taking 'something moves when motion is present' and 'if
something moves motion is present' as convertible [statements
(*antistrephonta*: cf. *An. Pr.* 25a6)] – from which it is inferred that, as
motion is, so too is moving – he attaches the consequent, producing a
933,1   conditional premise as follows: if motion is made of indivisibles, so
too is moving made of indivisibles. But this in fact will be proved to
be impossible; it will follow, accordingly, that neither is motion
made of indivisibles, and when this is posited, it will be
demonstrated simultaneously that neither is magnitude made of
5   indivisibles, which was demonstrated earlier in its own right. Here
it was demonstrated in addition that motion and magnitude are
alike in the respect that they are not made of indivisibles. He took,
then, the conversion [or convertibility (*antistrophê*)] of motion being
present and moving as self-evident: for it is not possible that
something move without motion being present or that there not be
10   moving when motion is present. He switched from motion to moving
because with the latter the absurdity that follows is more evident,
[namely,] that it [i.e. moving] is made of indivisibles,[15] than with
motion itself. From the difference between moving and having
moved arises, as we shall learn, the demonstration which denies
that motion is composed of partless things.

Taking it, then, on the basis of the illustration, that the thing O
15   moving the magnitude A moved [with] the motion that has D as its
label, he proves that it is not possible that moving[16] [with] one of
these motions be indivisible, assuming in advance that it is
impossible for 'something moving from somewhere to somewhere
simultaneously to move' toward that to which it is moving 'and to
have moved', and be there [at the place] toward which it was
moving. For it is 'impossible' for something walking to Thebes
'simultaneously to walk' to Thebes and to be in Thebes, when it is
walking. Rather, walking to Thebes occurs before being in Thebes.
20   Assuming this in advance, he switches once more to the illustration,
and says that O was moving [over] the partless line or extent A, to
which the motion D was present; for O was assumed to move

---

[15] Reading *to* with FM, as suggested by Leendert Westerink; Diels read *tôi*, i.e. 'in
being made of indivisibles'.
[16] Reading *kineisthai* with MSS rather than *kinoumenon* with Aldine, followed by
Diels, which does not accord with the argument here.

according to the motion D on the [line] A. It is necessary, accordingly, either that it traverse earlier and have traversed later, or that it simultaneously traverse and have traversed; for there is nothing between [these alternatives]. But if, traversing earlier, it 25 has traversed later, the motion would be divisible, which indeed is confirmed by the fact that when it was traversing, it was neither at rest as though it had not yet begun to move (for it is not in the place from which it was moving), nor is it already [in a state of] having traversed (for it would no longer be moving toward something), but it was in between the [place] from which and the [place] to which.[17] If the motion is in between and has not yet entirely occurred, it would be partitionable and divisible. This follows from [the fact 30 that] what is traversing earlier has traversed later, since every motion that is from somewhere to somewhere has portions, and one 934,1 [portion] of it has occurred, while another is about to occur. If someone should say that it simultaneously traverses and has traversed, then the thing walking will already have walked and have moved when it is walking, and be in that place toward which it is moving. What could be more impossible than this?

Then he adds something which indeed it was likely someone 5 would say, that O moves [with] the motion DEF over the whole magnitude ABC that is composed of partless things, though it is not itself partless, and that it is no longer moving [with] each of these partless motions but has moved [with each]; for thus the simultaneous moving and having moved will no longer seem to follow. For it will be moving over the whole [line], but no longer will 10 it be moving also along each of those indivisible portions. Having posited this, [Aristotle] draws absurd consequences by means of the following consideration: the motion will be composed 'not of motions [*kinêseis*] but of moves [*kinêmata*]', that is of the limits of motion, and of 'having moved'. For the motion DEF will not have the portions D and E and F as motions, for these [portions] were 15 supposed to have moved, not to move. Similarly, the portions of moving [*kineisthai*] too will be, not a moving, but rather a having moved, and moving will be divided into portions which are not a moving, but rather a having moved.

He adds yet another absurdity that follows upon this hypothesis, [namely] that something has moved that was not previously moving, for example, that something has walked that did not previously walk. For it is posited that O moves [with] the motion DEF over the 20 magnitude ABC, but it moves neither over A (for it has moved over it), nor over B, nor, likewise, over C. It will, consequently, have moved [with] the whole motion without previously moving [with] it.

---

[17] Diels closes the parenthesis here.

That he did not pose this objection as an entirely implausible
[possibility] is indicated by the fact that, although [Aristotle] both
25 posed it and thoroughly resolved it, nevertheless the Epicureans [fr.
278 Usener], who came later, say that motion occurs this way: for
they say that magnitude, motion and time are [all] made of partless
things, and that a moving thing does move over the whole magnitude that is constructed of partless things, but does not move along
each of the partless things in it; rather, it has moved [along the
partless things], because if it were posited that what is moving over
30 the whole [magnitude] were moving over these [partless things] as
well, they would be divisible.

He adds yet another absurdity to the argument that says that O
moves [with] the motion DEF over the magnitude ABC that is
composed of partless things but is not itself partless, yet it does not
move as well over each of the partless things ABC but rather has
935,1 moved [over them]. The refutation is evident. For, he says, if it is
necessary that what is so constituted [*or* of such a nature: *pephukos*]
as to move, when it is so constituted, either be at rest or move, then,
when O moves continuously [over] the partitionable [line] ABC, but
does not move in A or B or C, as they say, it would be at rest [in
5   them]. But it was supposed that it moved over the magnitude ABC
continuously. However, if it is continuously at rest in all the portions
ABC, it would also be at rest in all of it. Thus, the same thing would
simultaneously be both moving and at rest in respect to the same
thing, continuously; what could be more impossible than this?

He adds to the argument yet a fourth absurdity that was men-
10   tioned here and there in what was previously proved, but is here set
forth in an individual argument. The indivisible [parts of the
motion] DEF are either motions or not motions. But, if they are
motions, there will be something that is not moving but is at rest,
although motion is present to it; for when O was on A, the motion D
was present to it, when on B, the [motion] E, and when on C, the
[motion] F. But if DEF are not motions, the motion will not be
15   composed of motions, nor have parts of itself that are motions. It is
obvious that he who says that magnitude, motion and time are [all]
composed of partless things would not admit this as an absurdity,
but [it is so], since it was proved from the other [arguments] that
motion is not composed of partless things; on the other hand,
against one who says that over the magnitudes A, B, and C, there is
20   no moving but [only] having moved, this argument too has reasonably been added.

232a18-22: 'It is likewise necessary for length and motion' to
'the [line] of [magnitude] A will also be divisible.'

[Time must be made of indivisibles or divisibles in the same way as magnitude and motion; for if magnitude is divisible, a thing at a given speed will move over a smaller magnitude in less time, and similarly if the time in which it moves over a line A is divisible, then in a smaller time it will move over a part of A, which is consequently divisible.]

Having proposed to prove that magnitude, motion and time are alike in respect to being made of divisibles or indivisibles, and having proved first that motion is just like magnitude, and that each 25 is made of divisible parts, he adds here that time too is necessarily like the magnitude over which a motion [occurs] and the motion in respect to being made of indivisibles or not. For if those are made of indivisibles, then time too will be made of nows that are indivisible. That the other of these [propositions] is true is obvious, and since it was indeed self-evident, he did not add it here, [namely] that if the 30 former are made of divisibles [as opposed to indivisibles], then time too will be made of divisibles. But, having proved this one, [namely], that if magnitude and motion are made of indivisibles, then time too 936,1 will be made of indivisibles, he proves from this also what it is here proposed to syllogize, [namely] that time is like motion also in respect to being composed of divisibles or indivisibles.

According to the reading that says, 'for if every one [feminine] is divisible', that is, if every line [*grammê*] or extent [*diastasis*: both 5 feminine nouns] is divisible (for it is Aristotle's custom to speak of magnitudes [a neuter noun] in the feminine because he draws lines [in his illustrations] rather than magnitudes) – if, accordingly, a line or motion is divisible, since something moving at a uniform speed will traverse a lesser magnitude or a lesser part of a motion in less time (for in the same time something moving at a uniform speed will not move both a greater and a lesser [amount] either of magnitude 10 or of motion),[18] time too will be co-divided with the magnitude or the motion. For that of which it is possible to take a lesser bit will be divisible. He reasonably drew [the conclusion], accordingly, that time too will be divisible. Having thus proved from magnitude that time is divisible, he proves from time that magnitude is divisible by means of the same method. But perhaps one ought to say that it is 15 from motion that time was proved divisible, but magnitude from time, in order that we may, in an unforced way, understand the [feminine adjective] 'every' as pertaining to [the feminine noun] 'motion'. For if every time is divisible, then that in which something moves [over] the apparently indivisible magnitude A will also be

---

[18] In Diels' text, the words 'something moving at a uniform speed' (*to isotakhôs kinoumenon*) fall outside the parenthesis.

divisible, so that something moving at a uniform speed will, in a
[time] less than the whole [time], traverse less than A. A, conse-
20   quently, is divisible, since it is possible to take a lesser part of it.

If, however, there should have been written, as is found in certain
of the manuscripts, 'for if every single one [masculine] is divisible',
[in this case] Alexander explains that if every time [masculine] is
divisible, then magnitude too will be co-divided with the divided
time. Aristotle, however, does not draw [this conclusion], but rather
25   that time too will be divisible. Alexander too, having paid attention
to this later, says: 'Having proved that, if magnitude is everywhere
divisible, time too will be divisible, he assumed the converse in turn,
in the [sentence], "And if the time in which it proceeds [along the
line] A is divisible, the [line] of [magnitude] A will also be divisible." '

Aspasius, however, knew still another divergence in the reading,
30   as follows: 'for if every one [masculine] is' – and then he adds [the
word], 'indivisible'; and he explains that 'if every time were
hypothesized to be indivisible, but in less [time] something of
937,1   uniform speed will traverse a lesser [magnitude], then time, which
was hypothesized to be indivisible, will also be divisible.' Aspasius
remarked nicely here that [Aristotle] not only proves the
consequence that if the one is divisible the other will be as well, but
also necessitates in addition that time and magnitude be divisible
5   just like motion. For the demonstration from things of uniform
speed necessitates their divisibility, as we shall learn next. For the
same things are proved by an illustration as well, in this way: let O
be a thing of uniform speed, and let the magnitude AE over which it
moves be divisible. And let it be supposed that the thing of uniform
speed will traverse a lesser [magnitude] in less time. Let it traverse
the [line] AE in the time FG. If, accordingly, it traverses less than
10   AE, for example the [line] AB, it will traverse it in the lesser time
FH; the time too, consequently, will have been cut. Similarly, if we
cut the magnitude AB, with O advancing less, we shall again cut the
time FH. The time will always be cut along with the magnitude. It is
also evident that if time is divisible, magnitude too will be divisible,
again on the supposition that something of uniform speed moves
15   less in less time. For let the time, to which [is applied the label] FG,
be divisible, and let O move at a uniform speed [along] the [line] AE
in the time FG. Therefore, it will move [along] the lesser [line] AB in
the lesser time FH. Both the [line] AE, consequently, and every
magnitude will be cut similarly to the time FG.[19] Having proved,

---

[19] This proof is not spelled out in Aristotle, and is either Simplicius' own addition,
or is derived from Aspasius; the latter is perhaps slightly more probable in view of
certain syntactical constructions that are not customary in Simplicius, e.g. the
Aristotelian formula *eph'hôi* to indicate that letters are applied as labels, *kineisthai*
with *tên* [sc. *grammên*] as direct object, and possibly the use of *temô* here for *diaireô*.

thus, from the motion at uniform speed of a moving thing, that time
and magnitude are divided similarly, he proves the same thing from   20
faster and slower motion, teaching that magnitudes are partitioned
by something that moves more slowly, and time by something [that
moves] faster.

232a23-b20: 'Since every magnitude is divisible into magnitudes' to 'the faster will traverse in less time.'

[Since magnitude, being continuous, is always divisible, a
faster thing will traverse (i) in the same time a greater
magnitude; (ii) the same magnitude in less time; (iii) some
greater magnitude in less time. Proofs (i) If A is faster than B,
it will move or change over the magnitude CD in the time FG,
while B will not have reached D. (iii) Suppose A and B are
travelling over a track CEHD. Say B (the slower) has (in the
time FG) reached E, and A (the faster) D. A will then have
traversed CH in less time (say, FK) than that in which it
traversed CD, and thus in less time than that in which B
traversed CE; and CH is greater than CE. (ii) Since, by the
preceding, A can traverse a greater interval LM in the lesser
time PR, while B traverses the lesser interval LN in a greater
time PX, A will traverse the lesser interval LN in the time PS,
which is less than the lesser time PR and hence less than the
greater time PX.]

Alexander says: 'Here in this [passage], after having proved that   25
time is just like magnitude and motion in respect to being made of
partless things or not, [Aristotle] proves that time is not composed of
partless things or of nows.' But [we may reply that] this too has
already been proved, when [Aristotle] said, 'Time too will be
divisible' [232a21], that is, every portion of time, for a portion of time   30
is time. Rather, as was said [above], after having proved from the   938,1
motion at uniform speed of a moving thing both that when time is
divisible to infinity so too are magnitude and, obviously, motion as
well, and that they are divided proportionally to one another, here
he proves the same things from faster and slower [things]; for [he
proves here] not only that time is divisible, but that magnitude is   5
too. And Alexander himself agrees on this. Aristotle says towards
the end of the passage that 'the faster will divide time, the slower,
length' [233a7]. He demonstrates[20] antecedently, however, in this

---

[20] Reading *apodeiknusin* with the Aldine and 2nd corrector of A, rather than
apodeiknunai with Diels, who suggests in the apparatus criticus that *dokei* or

[passage] that time is divisible and, along with time, magnitude, which is what Alexander looked to. This is why Aristotle also added:

10 'It is simultaneously obvious that every magnitude is continuous' [233a10]. In the illustration, however, he proves both in parallel. For the proof of this, he assumes in advance that 'it is necessary that the faster thing move a greater [amount] in an equal time and more in less [time]' [232a25].[21] He proves these things both through the definition of faster (for we say that something that moves a greater

15 [amount] in an equal time and more in less time is faster), and from the fact that [the propositions that] in an equal time a thing that moves faster moves more and a thing [that moves] slower [moves] less follow from [the premise that] every magnitude is divisible. If it were not [the case] that every magnitude is divisible, it would not be possible for a slower thing always to move less than a faster thing in an equal time. For both a faster thing and a slower thing traverse an

20 atomic and a partless [magnitude] in the same time, for, if [the slower thing traverses it] in more [time], then in an equal [time] it will traverse something less than a partless thing. Therefore the Epicureans [fr. 277 Usener] maintain that all things move at a uniform speed through the partless [amounts], so that their atoms may not, by being divided, no longer be atoms. From the definition of a faster thing it is obvious that it is necessary for every magnitude to be divisible into magnitudes by virtue of faster and slower being

25 such [as they are], since this is the definition of them – unless, indeed, the faster and slower are such [as they are] on account of this, [that is], since every magnitude is divisible into forever divisibles.

Having thus concisely proved, accordingly, that a faster thing moves a greater [amount] in an equal time, he next proves the same thing through an illustration as well. For he posits A as a thing moving faster, and B as a slower, and assuming it in advance as

30 self-evident that what changes earlier than the slower thing is faster, he draws by way of illustration the time FG in which the faster thing changes and [in turn] the magnitude CD [along] which it changes in the time FG. Since, accordingly, in the time in which

939,1 the faster thing, A, has changed from C to D, the slower thing, B, does not yet reach D, but is left behind (since the slower thing is that

---

something similar has fallen out. But Simplicius is stating that Aristotle really did prove that the divisibility of magnitude is a corollary of that of time (cf. the clause beginning with *dio* = 'this is why').

[21] The text of Aristotle, according to the majority of the manuscripts, reads; 'it is necessary that the faster thing move a greater [amount] in an equal time and an equal [amount] in less [time] and more in less [time].'

which changes later), it is obvious that in an equal time the faster thing traversed more than the slower thing, if indeed the one [traverses] the whole [line] CD, while the other is left behind. Then he proves that the faster thing will also traverse a greater [amount] in less time – not greater by any amount whatsoever, for this is 5 false, but somewhat greater. For, with the same illustrations, since A has turned up at D in the time FG, and the slower thing, B, was left behind further in than D, at E, perchance, and there is some magnitude between D and E that is itself divisible, let it be divided at H. The [line] CH is greater than CE and less than CD. Since, 10 accordingly, A has moved [over] the line CD in the entire time FG, it will have moved [over] the [line] CH in less [time], for that [line] is less than the whole [line] CD. Let FK be the time that is less than the time FG. But in fact B was supposed to have moved [over] the line CE, which was less than CH, in the entire time FG. Consequently, B has moved in more time [over] a lesser line than CH, [over] which A has moved in the time FK, which is less than [the 15 time] FG. Consequently, the second [point] has also been proved, that the faster thing will traverse a greater interval in less time.

It might also have been proved readily in this way: for if the faster thing should have a speed in the ratio of three to one with respect to the slower, then, while the faster moves three feet, the slower will, 20 in an equal time, move one foot, so that the faster will move two feet, which is greater than one foot, in less time. But Aristotle seems to prove this for good measure [as a basis] for assuming that if a faster thing moves more in less [time], all the more will it move an equal [amount] in less [time]. For he will also make additional use of this [proposition] for the [argument from] division [below], and he proves 25 this same [proposition] for a third time as well by means of an illustration, saying that it has become 'evident' [232b5] from what has already been proved. This also could have been proved by means of the same illustration, but he altered the letters in order, I imagine, to exercise his readers variously, and [thereby] made the argument less clear. For he takes [the magnitude] LM instead of the magnitude CH, and LN instead of CE, while as the time in which A, 940,1 that is the thing moving faster, moves [over] the greater interval – CH or rather LM – [he takes] PR, which is analogous to FK. [As the time] in which it moves [over] the interval that is smaller than LM, [that is] CE or rather LN, [he takes] PS, which is obviously less than FK or rather PR. He also takes PX as the time in which the slower 5 thing B was supposed to have moved [over] the [line] CE, which he now has called LN. This was the [time] FG [and is] greater than PR, if PR, at all events, is analogous to FK. On [the basis of] such an illustration, then, he says that since it has been proved that the faster thing will traverse in the lesser time FK or rather PR the

greater [line], that is CH or rather LM, which the slower will
10   traverse in the greater [time] FG [or rather PX],[22] and a faster thing
assumed to be moving at a uniform speed, itself and in its own right,
will traverse in more time a [line] greater than the lesser line, for
example LM [which is greater] than LN, the time PR, in which it
traverses the [line] LM, would be more than [the time] PS, in which
[it traverses the line] LN. If, accordingly, the slower thing traverses
15   the [line] LN in the time PX, and the faster in [the time] PS, while
PS is less than PR and PR is less than PX, and that which is less
than a lesser thing is itself a lesser thing still, then it is obvious that
the faster thing will traverse the same [magnitude] LM, that is, an
equal [magnitude], in less time than the slower thing.

This [proposition], that a thing moving faster moves [over] an
equal interval in less time than a slower thing, he also proves in a
20   different way by [logical] division, as follows: if everything moving
[over] an interval equal to some [other] moves in 'an equal time' or in
more [time] or in less [time], and there is nothing besides these
[possibilities], a faster thing too, consequently, moving [over] an
interval equal to a slower['s] will move [over it] in more time or in an
equal [time] or in less [time]. But a thing that moves [over] an equal
or the same interval in more time is slower (for this is the definition
25   of a slower thing), and that which [moves] in an equal [time] is of the
same speed, while the faster is neither of the same speed as that
than which it is faster, nor still more is it slower. Thus, it will
traverse an equal [line] neither in an equal time nor in more [time];
941,1   in less [time], consequently. It is necessary as well to carry over the
definitions of a slower-moving thing from the faster, since Aristotle
himself uses not only the definition of the faster but also that of the
slower. If, accordingly, that which moves more in an equal time is
faster, then that which moves less in an equal time is slower; and if
that which moves a larger interval in less time is faster, then that
5   which moves a lesser in more time is slower; similarly, if that which
moves an equal interval in less time is faster, then that which moves
an equal [interval] in more time is slower.

232b20-233a12: 'Since every motion is in time' to 'time and
magnitude are divided in [the same] and equal divisions.'

[Since a thing may move faster or slower in any time, time is
continuous (= forever divisible). For the slower divides the
distance, the faster divides the time. Thus if a slower thing B
traverses the magnitude CD in the time FG, the faster thing A

---

[22] Adopting Diels' conjecture (in the apparatus criticus), *tôi <PX êtoi tôi> ZH.*

will traverse the same magnitude CD in the shorter time FH
(by the preceding argument). In the time FH, in turn, B will
traverse the lesser magnitude CK; but A will traverse CK in
still less time, and so on; thus every time and magnitude is
continuous (= forever divisible).]

Having proved that a faster thing moves an equal interval in less   10
time, and a slower [moves] a lesser [interval] in an equal [time], he
makes use of these [propositions] in addition to prove that time is
not composed of nows that are partless, but rather is continuous and
divisible into forever divisibles, and likewise magnitude as well, and
the faster thing divides time, while the slower [divides] magnitude.
Lest someone say that not every motion is in time and that the   15
faster and the slower are not in every motion, and [that] if this [is
the case], then not every time will be divided by the faster, he
mentions these things as having been proved and agreed upon in the
previous [passages], [namely] that every motion occurs in time and
motion [occurs] in every time, since time has been proved to be a
[property] of motion [cf. 219a8]. Thus wherever there is motion,
there is also time, and where there is time, there there is also   20
motion. He assumes in addition to this that every moving thing can
move faster and slower, and Alexander poses well the puzzle [of]
how moving faster and slower is true in the case of the revolving
body [of the heavens], which moves evenly [cf. 223b19]. He resolves
[the puzzle] first in a weak way, I think, claiming that Aristotle did
not say that the same thing [actually] moves faster and slower, but   25
that it can [so] move, which, he says, is a property of the revolving
[body] too because it moves thus [i.e. uniformly] according to its own
will, and not under necessity or as though moving also in a different
way has been prevented by someone. 'For good men,' he says, 'do not   942,1
do good [deeds] by necessity, even if they always do them, but they
have the power of [doing] the opposite as well.' But he ought, I think,
to have noted first that good men are said to have the power of
[doing] the contrary, because sometimes they also [actually] act
according to it [i.e. the contrary of the good], but how could things
that are heavenly and eternal have a power that never emerges into
actuality? Next, [he ought to have noted] that necessity is of two   5
kinds, one [kind] being more divine, according to which it is
necessary that god be good and that untainted souls never be
corrupted, the other [kind] being forcible, according to which bad
men too are constrained by laws not to err. Power too is of two kinds,
one [kind] being perfect, the other imperfect and rather in potential.
Heavenly things, accordingly, both are under the more divine   10
necessity of always moving in the same way and have a will that is
determined in the good and completely perfect and pure of any

ambivalent power; for divine souls do not possess the good in a way
similar to human souls, but rather the latter [have the good as] a
finite thing and one that sometimes changes into its contrary, while
the former [have it as] an infinitely powerful thing and forever
settled in the same condition. Next, [Alexander] solves [the puzzle]
well by [the notion of] power [*dunamis*]: he says that a motion has
15   the same interval as the magnitude over which it occurs;
accordingly, both a thing moving faster and a slower thing have the
power to accomplish the interval of a motion that occurs over a stade
[approx. 607 ft.] or over one section of the zodiac, but the one [has
the power to do so] in less time, the other in more. And he brings in
Eudemus [fr. 101 Wehrli] too, in the fourth [book] of his *Physics*, as
20   demonstrating what was stated [by Aristotle], thus: for [Eudemus]
himself says, 'since every moving thing is in time, in every time
there is motion, and speed and slowness attend every motion, it is
necessary that all the said things [i.e. time, motion and magnitude]
be partitioned. For it is possible to assume something moving faster
and slower upon the same magnitude, even if it is impossible for the
same thing to move now faster and now slower.'
25   Assuming that, because motion occurs in every time and every
motion has a faster and slower, faster and slower motion occurs in
every time, [Aristotle] adds that 'these things being so, it is
necessary also that time be continuous'. And he specifies that the
continuous is that which is 'divisible into forever divisibles'. He then
30   adds that if the continuous is such, it is necessary that time be
continuous, since in every time there is faster and slower motion.
And he proves that both time and magnitude are forever divided,
943,1 and that time is divided by the faster, magnitude by the slower. For,
taking it as previously agreed upon that the faster will traverse an
equal interval in less time than the slower, and that the slower [will
traverse] in an equal time a lesser interval than the faster, and
again illustrating the faster-moving thing as A, the slower as B, and
5   the magnitude [over] which the slower thing moves in the time FG
as CD, he says that the thing moving faster will move [over the line]
CD in less than the time FG. Let this [time] be FH. Accordingly,
since the faster has traversed [the line] CD in the time FH, it is
obvious that the slower will traverse less in the same time because
10   motion occurs in every time, and that on what[ever] the faster thing
moves the slower too can move. Accordingly, let CK be the
magnitude which the slower thing, B, has traversed in the time FH.
Again, since the slower has traversed the magnitude CK in the time
FH, the faster will traverse it in less [time]. The time FH,
accordingly, will again be divided. But in the time, again, in which
15   the faster has traversed the magnitude CK, the slower will traverse,
again, less than the magnitude CK. And thus time will forever be

divided by the faster, while magnitude [will be divided] by the slower, because motion occurs in every time, and every motion can be both faster and slower. For what has been proved follows of necessity from these suppositions, by which it is obvious that every 20 time is divisible; for in any [time] that is taken, the slower-moving thing will be traversing some magnitude, which the faster-moving thing will traverse in less time, in which time the slower-moving thing will traverse a lesser magnitude. And thus, with the faster and the slower forever converted, time is divided by the faster to infinity, and magnitude by the slower. Here he either calls the 25 alternation from the faster to the slower and from the slower to the faster 'conversion', or else it is possible to take conversion in the following [sense]: if the faster divides time, then the slower also [divides] magnitude, and [conversely] if the slower [divides] magnitude, then the faster also [divides] time [cf. *An. Pr.* 25a6]. These things have been proved from those that were proved previously, [namely] that a faster thing moves an equal [magnitude] 30 in less time, and a slower thing [moves] a lesser [magnitude] in an equal time. Having proved [these things] concerning time, [Aristotle] reasonably adds that it has become obvious from what has been said that magnitude too is simultaneously continuous, which is the same as saying that it is divisible to infinity, because both the time and magnitude are divided [into] the same and numerically equal divisions.[23]

It is reasonable that some find fault with the demonstration even 944,1 here as circular and proving itself by means of itself, since he proves that the faster moves a greater [magnitude] in an equal time and more in less [time] and an equal [magnitude] in a lesser [time] from [the fact] that every magnitude is divisible into magnitudes; and from this [he proves] that time is forever divisible, and from this 5 that magnitude is such. For he will seem to be saying: since magnitude is forever divisible, time too is forever divisible, and since time is, magnitude is too; and in short, since magnitude is forever divisible, magnitude is forever divisible.[24] But it must be said that first of all and apart from the assumption in advance that every magnitude is divisible into magnitudes, he proves from the 10 definition of a faster thing and from the illustration what was proposed [for proof]; and then, that he proves antecedently here that time is divisible, and demonstrates simultaneously with this that magnitude is too. In the third place, moreover, it must be said that through such a demonstration Aristotle indicated that because magnitude and time are continuous in a naturally similar way

---

[23] Reading *tas* instead of *pas* (Diels), as suggested by Cecilia Trifogli.
[24] Reading *esti* with F, instead of *ésti* with Diels (or *estai* with CM).

[*homophuôs*] they are also demonstrated from one another in such a
15 way as to seem responsible for one another's division to infinity. But
[we may reply that] the procedure of the whole argument is of such a
kind: having proved earlier from the motion at a uniform speed of a
moving thing that neither magnitude nor motion nor time is
composed of partless things, but that they are continuous and
divisible into forever divisibles, and having proposed to prove this
20 very same thing from a faster and slower moving thing as well, first
he proves it concisely by the potential [syllogism] thus, saying that
unless every magnitude were divisible into magnitudes, that of time
and that of length and that of motion, the faster and the slower
would not be such as we say they are in defining [them], saying that
the the thing moving a greater [magnitude] in an equal time and
25 more in less time is faster. Now, since magnitudes are divisible, the
faster and the slower are, therefore, the divisors [*diairetika*] of
them, the faster of time, the slower of length. There is also a
confirmation on the basis of [cor]relatives [*apo tôn pros ti*; cf. *Cat.*
11b20]. For since it is self-evident that the faster and the slower are
both in existing things and are divisors of time and magnitude,[25] it
945,1 is necessary that the divisible exist, since there exists a divisor. But
there exists a divisor, thus also a divisible, as has also been proved.
Then, having supported by means of the illustration [the fact] that
faster and slower are such as they are said [to be], he next uses them
for the division of time and length, showing the character of the
5 division to infinity by means of the illustration.
It is necessary to attend to [the fact] that he proved here that time
and magnitude are divisible by means of motion. For the faster and
the slower are differences within motion, because 'it is possible for
every moving thing to move both faster and slower', as he himself
10 says, over the same interval of motion, obviously, which is [in turn]
taken from the interval over which [the thing] moves. It would not
have been possible for something both faster and slower to move
[over] the same interval of motion, unless the magnitude of the
motion had also been divisible. The three magnitudes, accordingly,
that of motion, of time, and of the length over which the motion
[occurs], are proved from one another to be similarly continuous and
15 divisible into forever divisibles. For if one [of them] is divisible, then
it is necessary that the rest be divided. For if length is divisible, it is
necessary that the motion over it also be divisible, and the time that
measures [the motion]; and if motion is divisible, then it is necessary
that both the length over which it occurs and the time that measures
20 it be divisible; and if time is divisible, then it is necessary that the

---

[25] Deleting *epeidê* 'since'; Diels indicates a lacuna, which he supplements in the
apparatus criticus; '<it is necessary that the magnitudes of time and length be
divisible as well,> since...'

motion measured by it be divisible and the magnitude over which
the motion [occurs]. Just as the [length] over which the motion
[occurs], if divided, divides the motion, so too the motion [divides]
the length over which it occurs.

233a13-21: 'Furthermore, it is evident also from the arguments
that are usually stated' to 'if by both, then the magnitude too
by both.'

[If time is continuous, so is magnitude, since a thing traverses
a lesser magnitude in less time, and vice versa; further, if
either is infinite, whether by extremes (i.e. in extent) or by
division, so is the other.]

Having proved by means of the illustration that magnitude too is 25
co-divided with the division of time, he says that the same thing is
proved also from other arguments that were customary, as it seems,
among those who discussed such things at the time. The argument
taken from things moving at a uniform speed was of the following
sort: if something moving at a uniform speed 'traverses in half the
time half, and always in less time less', and less according to the
same ratio, it follows that, since the time is divisible to infinity, the 30
magnitude too over which the motion [occurs] is similarly divisible 946,1
to infinity, for there are the same divisions of either. If they are
similar to one another, it is obvious that if the one is infinite, the
other will also be infinite. Since the infinite in continuous things is
of two kinds, or rather can be thought of in two ways, either 'by 5
extremes' (what has no extremes, but rather is untraversable, is
infinite by [virtue of its] extremes) or 'by division' by virtue of the
fact that [any] portion that is taken is always divisible, or according
to both [senses], in whatever way time possesses the infinite, so too
will magnitude possess it. But it is obvious that just as time, being
continuous, is divisible to infinity, so too is magnitude, since it too is 10
continuous: every continuous thing is divisible to infinity. But since
time is also infinite by extremes, having neither a beginning nor a
limit, as is proved in the last [book] of this treatise [263a11], how is
it true that magnitude too is infinite in this way? For the very
opposite has been proved in the third [book] of this treatise
[206a7ff.], [namely] that no magnitude is infinite by [virtue of its] 15
extremes. But just as time, then, is of two kinds, the one infinite by
[virtue of its] extremes, measuring a beginningless and endless
revolution, the other, limited by extremes, [and] possessing the
infinite in division alone, like the particular kind that measures
motions on a straight [line], so too magnitude is of two kinds, on the

one hand infinite by extremes, not as being indefinite [in length
20 (*aoristos*)] such as he rejected in the third book, but as circular and
having neither beginning nor limit, and on the other hand having
both beginning and limit, like things in [the process of] coming to be,
but which are divisible to infinity because they are continuous. Both
time as a whole and the revolving body [of the heavens] partake of
both infinities. For they are both beginningless and endless, and
25 since they are continuous they are divisible to infinity. It occurred to
me, indeed, both to pose this puzzle in this way and to solve it, since
I have found it in none of the commentators, but if anyone should
address [it] in a more plausible way, he wins as a friend rather than
an enemy [cf. Plato *Timaeus* 54A5]. It is obvious that even though
[Aristotle] mentioned time and magnitude alone in this [passage], it
30 is necessary to understand the same things concerning motion, for
just as time is, so too is motion of which time is the measure.

947,1   **233a21-31**: 'Therefore too the argument of Zeno assumes
something false' to 'and to touch an infinite [number of things]
by an infinite, and not a finite, [number of times].'

[Zeno is wrong to assume that a moving thing must first go half
its journey, and before (or after) that half of the half, and so on,
so that it has to traverse or touch an infinite number of things,
in a finite time. For the time and the distance are no more (and
no less) infinite than each other. Both are infinitely divisible,
neither is infinite by extremes (i.e. in length).]

Having proved that the infinite exists in a similar way both in time
and in magnitude, he uses this in addition to solve one of the
5   arguments posed by Zeno the Eleatic with regard to the
non-existence of motion. Zeno's argument is of this sort: If there is
motion, it is possible in a finite time to traverse an infinite [number
of things], touching each of them; but in fact this is impossible;
consequently, there is no motion. And he proved the conditional
premise by using the division of magnitudes to infinity: for if every
magnitude is divisible into an infinite [number of things], it would
10 also be composed of an infinite [number of things]; thus, a thing that
is moving [over] and traversing any magnitude whatever would
move [over] and traverse an infinity and would touch an infinite
[number of things] in a finite time, in which [time] it will traverse
the whole finite [magnitude]. [Aristotle] says 'touch the infinite
[number of things] one by one', since something can seem to have
traversed an infinite [number of things] by skipping over them.
15 Thus did [Zeno] prove the conditional premise. He proves the

additional premise that says, 'But, in fact, it is impossible to traverse an infinite [number of things] and to touch them in a finite time', from the fact that the infinite is untraversable and from the fact that it is not possible to touch an infinite [number of things] in a finite time, if, at all events, the moving thing touches the parts of the underlying thing in one [moment] of time after another. But he said that it was impossible to touch each of an infinite [number of things] because he who touches counts, as it were; and it is impossible to 20 count an infinite [number of things]. But Aristotle impugns the additional premise: for it is not impossible to traverse an infinite [number of things] in a finite time. For, since the infinite is spoken of in two ways, on the one hand in respect to extremes, and on the other hand by division, it is impossible for something to traverse in a finite time the infinite in respect to quantity, the parts of which are numerically infinite in actuality, because the magnitude too is 25 infinite, but it is possible [to traverse in a finite time] the infinite according to division and potentially, not actually [so]. For the time taken as finite is also infinite in this way, in respect to division: not, accordingly, the infinite in a finite time, but rather in a similarly infinite [time]. And the moving thing will touch, accordingly, a potentially infinite [number of things], but not portions actually divided, one by one, of the underlying thing, in parts of time infinite 30 in a similar way as in [the case of] its own [parts]. For in fact the moving thing too, being continuous, is divisible to infinity.

Zeno's paradox, accordingly, arose by way of the equivocality [or homonymy (*homonumia*; cf. *Cat.* 1a1-5)] of the infinite and by substituting the infinite by extremes for the infinite according to 948,1 division. For it would be impossible to traverse this infinity in a finite time and to touch a thus infinite [number of things], but traversing a magnitude infinite according to division in a time infinite according to division is in no way absurd. Zeno took magnitude as actually divided into an infinite [number of things], but he did not further [take] time [that way], even though it is 5 similar to magnitude. But [Aristotle] here solved Zeno's puzzle in this way, by saying that there is nothing absurd in traversing things infinite in one way in a time infinite in a similar way, while in the last [book] of this treatise he says that this solution 'is sufficient with respect to him who posed [the puzzle] (for what was posed was whether it is possible to traverse entirely an infinite [number of things] in a finite time)' [263a15-17], but with respect to the [real] 10 issue and the truth it is not sufficient: for neither does time possess the infinite actually, but rather potentially. Therefore he solves it there more accurately, by proving that there is not in continuous things an infinite [number of things] in actuality, but rather potentially. He says: to someone who asks, accordingly, whether it is

possible to traverse entirely an infinite [number of things] either in time or in a length, it must be said that there is a way in which it is possible, and a way in which it is not possible: for it is not possible
15   [to traverse] an actually existing infinite [number of things], but it is possible [to traverse] a potentially [infinite number of things].

233a31-b7: 'Nor indeed is it possible [to traverse] the infinite in a finite [time]' to 'for it will be divided into equal [parts] as is the magnitude too.'

[It is impossible to traverse an infinite magnitude in a finite time or a finite magnitude in an infinite time. For if AB is a finite magnitude, and C an infinite time, in a finite portion CD of C a thing will traverse a finite portion BE of AB. Now, a certain (finite) number of BE's will measure off (whether evenly or not) AB; that same multiple of the finite temporal interval CD will measure off the whole time C, which is therefore not infinite after all.]

Having proved that time and magnitude are similarly continuous and similarly divisible to infinity, and having thoroughly resolved
20   the argument of Zeno on the basis of this similarity, he adds something as following upon the things that have been previously proved and proved from those things, [namely] that it is impossible for something to move either [over] an infinite interval, even on the hypothesis that it is in a finite time, or [over] a finite [interval] in an infinite time. He takes the infinite here as the properly infinite, in actuality, [the one] which, indeed, he called infinite by extremes. He
25   produces the proof by impossibility through an illustration, taking a finite magnitude, to which AB [is applied as label (cf. n. 60)], and an infinite time, to which C [is applied], and separating off a finite bit of time, to which CD [is applied], in which the moving thing will traverse a bit of the magnitude AB, for example, BE; for it will move over some portion of it in the time CD, since it was moving [over] the
30   whole AB in the infinite time C. However many times [greater], then, the magnitude AB is than the individual portion BE, so many times greater too will the whole time be, in which the moving thing will traverse evenly and at a uniform speed the whole magnitude AB, than the time CD, in which it moves over the magnitude BE.
949,1   But the time CD is finite; consequently, the time C, in which it will traverse the finite magnitude AB, is also finite, and not infinite. For the time C [is divided] by CD into just as many equal [parts] as the equal [parts] into which the magnitude AB is divided by BE. But the
5   magnitude AB and the time CD were finite; thus, the time too, just

as the magnitude, will be divided into a [number of parts] finite both in magnitude and in number. Such, then, is the demonstration.

In the midst of resolving the objection that says that the magnitude BE does not strictly measure off the whole of AB, in such a way that one may truly say that AB is three times greater, perchance, than BE, or four times or so-and-so many times (for it is possible both to fall short and to exceed in a measurement), [Aristotle] says, accordingly, that it makes no difference whether it falls short or exceeds, provided only that it traverses a [magnitude] equal to the magnitude BE in a [time] equal to the time CD. For, just as the magnitude BE exhausts AB, whether it exceeds it or falls short (for that which falls short, falls short by some part [i.e. fraction] of itself, and that which exceeds similarly exceeds by some quantity), so too the finite time CD will exhaust the time C, whether it falls short or overshoots, becoming as many times greater than itself as the magnitude AB is [greater] than the magnitude BE,[26] whether precisely, or falling short by some part of itself,[27] or overshooting. For the magnitude AB will be measured off by the [magnitude] BE and the time C by the [time] CD simultaneously by virtue of being divided into [parts] equal in number.

It is necessary to attend [to the fact] that when [Aristotle] says 'this will either measure off [the magnitude] to which AB [is applied as label] or will fall short or will exceed,' he is using [the term] 'measure off' properly. For to measure off is for the thing multiplied to be made precisely equal to the thing measured off. But when he said 'this measures off the whole [magnitude]', he used [the term] 'measure off [*katametreô*]' instead of 'exhaust [*dapanaô*]', For whether it falls short or overshoots, the time C, in which [the thing] moved [over] the whole magnitude AB, will also be divided into [parts] numerically equal, that is, into just so many [parts], as those into which the finite magnitude was divided. Thus, it will follow that the [time] is finite. For however many times greater it may be than the finite [time], whether exactly or with some portion [i.e. fraction], it will be finite.

Given that the reading at the end of this passage is reported in two ways, if what may have been written is, 'for it will be divided into equal [parts] as is the magnitude too', then what is being said would be clear, [namely] that the time too will be divided into numerically equal [parts], as is the magnitude too. But if, as Alexander and Aspasius and Themistius [187,28-9] write, [the reading is] 'for it will be divided into equal [parts] – magnitude too', then [Aristotle] would be saying: it [i.e. time] will be divided into a

[26] Following the reading of MSS CM; Diels, following the MSS FA, reverses AB and BE.

[27] Reading *heautou*, with A, instead of *autou*, 'of it'.

[number of parts] equal to those into which magnitude too [is divided].[28] For he is speaking about time, having proved that this [i.e. time] is not infinite.

233b7-15: 'Further, if not every magnitude' to 'the demonstration is the same also if the length is infinite and the time is finite.'

[Further proof that the time is finite: if there is a magnitude BE that is traversable in a finite time and that measures off the whole magnitude in question, and if the moving thing traverses an equal magnitude in an equal time, then the whole time will be finite. There does exist such a magnitude BE, since we can take a time bounded on one side in which the moving thing traverses some magnitude, and the thing will traverse a part of this magnitude in a lesser time than the whole time. Similarly for an infinite magnitude and a finite time.]

Having earlier taken the magnitude BE as rather indefinite and as
10   not strictly measuring off the [magnitude] AB, but sometimes exceeding or falling short, with the motion on it measured by the finite time CD, both of which [assumptions] seem subject to certain objections, throughout this [i.e. the present] argument he separates out a [magnitude] BE of such a sort as to measure off the whole, and assumes by hypothesis that the thing moving through it traverses it
15   in a finite time; and he proves that even if the [magnitude] BE is hypothesized to measure off from the beginning the [magnitude] AB, if the time CD is taken as finite the whole time too will be proved to be finite. For if BE is taken as the tenth part of AB, a thing moving at a uniform speed will move [over] the ten parts that are equal to BE in ten [stretches of] time equal to CD. The ten finite parts of the time, added together, make the whole [time] finite. He
20   then proves that it is not possible to traverse the magnitude BE, which is finite and a part of the [magnitude] AB, in an infinite time, since [the moving thing] traverses the whole [magnitude] AB in an infinite time. For, hypothesizing that the time C is finite [i.e. bounded] on one side or the other, and taking it as self-evident that something moves [over] a portion in less time than [over] the whole, he proves that the time in which something moves [over] BE is finite
25   on each side. For, starting from the same beginning, which is finite

---

[28] The issue concerns the presence or absence of the conjunction *hôs*; without it, the last clause appears to say that magnitude too is divided into an equal number of parts, which Simplicius regards as an inconsequential conclusion to the argument; hence, he offers a forced reading of the syntax.

[i.e. bounded], [the moving thing] will delimit [the time] on the other side as well; for if this is not [the case], it will no longer be traversing the part in less time than the whole, but rather in an equal [time].

He took the time C[29] as finite on one side or the other, so that he might separate out that [time, i.e. CD], in which [the moving thing] will traverse the [magnitude] BE, as finite on each side. This [i.e.   951,1 that the time is finite on one side] would easily have been conceded to him; for if it is not possible to traverse the [magnitude] BE in a [time] that is infinite on one side, [it would be] all the more impossible in [a time] that is infinite on both sides. But if he took it as self- evident that it will traverse the part in less time, and it is necessary that the finite be less than the infinite, what further   5 argument is required for [the conclusion] that it will traverse the magnitude BE in a finite time, unless, after all, the infinite on one side, although it too is infinite, is less than the infinite on both sides? But if this [is the case], why did he hypothesize that the time in which the moving thing will traverse the magnitude AB was infinite on one side? For if he hypothesized that it was infinite on both sides, nothing prevented [the moving thing] from traversing the [magnitude] BE, which is a part of AB, in a [time] less than the   10 [time], infinite on both sides, in which it will traverse AB, [that is], in a [time] that is infinite on one side and which itself is a part of the [time] that is infinite on both sides. But [we may reply], if it is necessary that the time be measured off by the [lesser] time just as the magnitude AB is measured off by the magnitude BE, it is impossible that the one [stretch of time] be infinite on both sides, and the other on one side. For an infinite time, being indefinite and indenumerable and measureless, neither measures off nor is   15 measured off. Having hypothesized, accordingly, that the [magnitude] BE measures off AB, and taking it that both the time and the magnitude will be divided into numerically equal [parts], he avoided such objections. For it is necessary that that which measures off the finite be finite, like the [magnitude] BE. And that which measures [it] off by a number of times equal to a finite [number], for example ten times or twenty times, cannot be infinite even on one side. For if   20 that which is twenty times greater than it is finite and defined and numbered, so much the more [will] the twentieth [part] of the twenty times [be so].

Having thus proved that it is impossible to move [over] a finite magnitude in an infinite time, it is possible, he says, using the same [arguments], to prove that neither is it possible to traverse an infinite [magnitude] in a finite time. For, if the time is hypothesized   25 to be finite, I shall separate out some portion of it, by which the

[29] Reading 'C', with the MSS; Diels emends to 'CD'.

whole [time] will be measured off, and I shall separate out some magnitude from the infinite magnitude as well, on which [the thing] moved in the time that was separated out. And since it moves an equal [magnitude] in an equal time, the time will be measured off by the time and the magnitude by the magnitude an equal number of
30   times. But that which is composed of [parts that are] finite both in magnitude and in number will be finite, not infinite. Themistius [188,13- 189,1] remarked well that 'even if it does not move [over] an equal interval in an equal time, but rather sometimes an equal
952,1 [interval] in more [time] and sometimes in less, with the finite interval divided even in this way into equal parts the time too will be co-divided into the same number of parts, no longer into [parts] equal to one another, but for [the purpose of] showing, at all events, that the [time] is finite, it suffices to show that it is made of portions that are finite both in number and in magnitude, even if the parts
5   should happen to be unequal (for it is necessary that what is made of portions that are finite both in number and in magnitude be finite); similarly and conversely, if the time is finite, and the interval should be hypothesized to be infinite. It is only possible to move [over] a finite interval in an infinite time, then, if [something] should move again and again [over] the same [interval] in a circle, as the divine bodies do (for motions that bend back on a straight [line] are
10   interrupted by a stop, as will be proved in the last [book] of this treatise [264a20-21]). But if a finite body will not traverse an infinite interval in a finite time, it is evident that neither would an infinite body, if we should hypothesize that it is moving, traverse a finite interval in a finite time. For whenever an infinite body traverses a
15   finite line, a finite line will also entirely traverse an infinite body. But this has been demonstrated to be impossible. And if, indeed, we should hypothesize both the interval and the moving body to be infinite, and only the time to be finite, it is necessary that the rest prove to be finite as well.'[30]

20   **233b15-32**: 'It is evident, accordingly, from what has been said' to 'it is evident, accordingly, that none of the continuous things is partless.'

[If a continuous thing were atomic (i.e. uncuttable), what is uncuttable would turn out to be divided: for if a faster thing traverses one and a half times the magnitude traversed by a slower in the same time, let the greater magnitude be divided

---

[30] Quoted from Themistius with variations; major additions by Simplicius are in parentheses.

into the atomic parts AB, BC, and CD, to which correspond the atomic times KL, LM, and MN, and let the lesser magnitude be divided into the two atomic parts EF and FG, and the time of the slower thing (= KN) likewise. It follows that the atomic time BC will be cut in half, and that the partless magnitude EF will be traversed in more than an atomic time. Hence no continuum is partless.]

Having proved that it is impossible for any of the continuous things to be composed of partless things – neither magnitude, nor motion, nor time – and that every continuous magnitude is divisible into forever divisibles (for it is necessary either that it be indivisible or   25 divisible, and if divisible, then either into indivisibles or into forever divisibles; but it is not indivisible, since it is a magnitude, at all events, nor is it divisible into indivisibles, for [in that case] an   953,1 indivisible will be touching an indivisible; thus into forever divisibles), and having proved also that time and magnitude possess the infinite and the finite in a similar way, from which he also solved the argument posed by Zeno concerning the non-existence of motion, he says here that since it has been proved by us that continuous things are not divided into atomic [parts], it is evident as well from   5 this that no portion of continuous things is atomic and indivisible, since continuous things are not divisible into atomic [parts]. The argument will prove also in the case of the portions themselves that it is impossible for them to be atomic: first, for a [portion] of magnitude, then also for a [portion] of time. This demonstration would differ from the earlier one, because that one proved that they are not indivisible, while this one proves antecedently that they are   10 divisible, and shows the necessity of division. And indeed, it is from the nature of a continuous thing that the portions of it were previously proved divisible, while here the portions themselves are demonstrated to be such in their own right. He says that part of a line is a line and part of a plane is a plane and a [part] of time is time, and, in general, part of a continuous thing is continuous; for   15 the argument here concerns portions in continuous things – that they are not indivisible, which he first proves for magnitude, and then also for time, using in addition the fact that the time and the magnitude on which a motion [occurs] are similar to one another. He proves it through a reduction to impossibility, syllogizing potentially [cf. n.6] according to the second hypothetical [figure of inference], as   20 follows: if a part of the magnitude is atomic, the atomic will be divided; for a part of the magnitude is divided; but in fact it is impossible for the atomic to be divided; consequently, the antecedent, that part of the magnitude is atomic, is also impossible.

He proves the conditional premise by using in addition the fact that
it is self-evident and assumed in advance that the faster and the
25   slower exist in every [stretch of] time, since there is motion in every
[stretch of] time, and the faster and the slower exist in every motion,
and further, indeed, the fact that in an equal time a faster-moving
thing moves [over] a greater interval than a slower-moving thing.
These things having been posited, he takes as a third [premise] the
fact that it is possible to take the motions of the faster thing and the
slower thing in some ratio, as [for example] that in the same time
30   the faster moves [over] twice [as large an interval] as the slower, or
[over] one and a half times [the interval]. Assuming these things in
advance, accordingly, he proves through an illustration what is
proposed [for demonstration], hypothesizing that the motion of the
faster is one and a half times greater in respect to that of the slower,
so that in the time in which the slower moves [over] a certain
magnitude, the faster moves one plus a half of that [magnitude]. Let
954,1   these two magnitudes, accordingly, be divided, he says: that – AD –
which is moved [upon] faster, into the three parts AB, BC and CD,
and that – EG – which [is moved upon] more slowly, into the two
[parts] EF and FG. In this way the ratio is one and a half. Since the
time is similar to the magnitude on which the motion [occurs], the
time KN, in which the faster thing moves [over] the [line] AD, will
5   itself also be divided, according to the hypothesis, into the three
times KL, LM and MN. For a thing moving at a uniform speed will
traverse an equal [magnitude] in an equal time. Since, accordingly,
the slower-moving thing in the same time moves [over] the [line] EF,
FG, which is made of two partless things, it will move [over] the
[line] EF, consequently, in half the time. Consequently, the time
also will be divided in two, just like the magnitude on which the
10   slower thing is moving. If the [time] that is composed of three atomic
[parts] should be divided in two, the middle atomic [part] would be
divided; for it is not otherwise possible for something made of three
equal [parts] to be divided into two equal [parts]. But the atoms are
equal, nor is one greater and another less, so that two [atoms may]
be equal to one. He reasonably drew [the consequence], accordingly,
that 'the atomic will, consequently, be divided', which is indeed
impossible. Also impossible, consequently, is that upon which this
followed [as a consequence], [namely] that a part of a continuous
15   thing is atomic. He draws also another absurd consequence, that '[a
moving thing] will not traverse a partless entity in an atomic [time]'.
For the slower-moving thing will traverse the magnitude EF,
according to this method [of argument], not in the atomic time KL,
which indeed is what was needed, but rather it will move [over] less
than the partless [magnitude] in a partless [time]. But what could
be less than a partless [magnitude]? And in fact the partless thing

will also be divided in the following way, since there is something   20
less than it: into both a lesser [part] and an excess [over that].
Having proposed to prove above all in this [passage] that the parts of
magnitudes are not atomic, because [the consequence] that atomic
[parts] are being divided, which indeed is impossible, follows upon
the [premise] that says that they are atomic, he proved that the
parts of time that were hypothesized to be atomic were divided, but
not yet that the [parts] of magnitude were, although he took
magnitudes to be one and a half times [larger], when he said, 'let the   25
faster, accordingly, proceed [over a magnitude of] one and a half
times in the same time'. [We may reply,] then, [that] it was indicated
by means of the parts of time that the parts of magnitude too will be
divided, although they were hypothesized to be atomic. For if, when
the time KN was cut in two at Q, the portion LM, which was
hypothesized to be atomic, was divided, the faster-moving thing will
traverse in the half time KQ half of the magnitude AD, and the   30
portion BC, which was hypothesized to be atomic, will be cut in two.

233b33- 234a7: 'It is necessary too that the now which is not by
virtue of something else [be divisible]' to 'because a continuous
thing is not made of partless things.'

[A 'now' (instant, or present instant) in the primary sense of
the term, is indivisible, and exists in every stretch of time; it is
the extreme limit of the past and of the future, and if these two
limits are one and the same, it must be indivisible. The two
limits cannot be consecutive, since then a continuous thing
(time) would be made up of partless limits.]

Having proved that the parts of continuous things are divisible,
[that is], both of time and of magnitude, and also, obviously, of   955,1
motion which is intermediate [*meson*] between them [i.e. partakes of
both], from the fact that, if they were not divisible, the atomic [part]
would be divided, which indeed is impossible, he next – since a now
too is something that belongs to time – articulates, reasonably, the
things [that obtain] concerning it. Since the 'now' is of two kinds, on
the one hand in the sense of a beginning and limit of time, which is
analogous to the point in a line, and on the other hand spoken of in   5
connection with the present time, which is distinguished in
opposition to the past and the future, he says that the one, [that is],
in the sense of a beginning and limit, which is also indivisible, is
called a now 'in its own right and primarily', while the other is not
[so called] in its own right or primarily, but rather insofar as it has
within itself the [one which is so] in its own right and primarily. For

10   in this way too, surely, something which is within the limit of what
     immediately surrounds [it] is said to be in a place in its own right
     and primarily, while something else is not [in a place] in this way, as
     [when something is said] to be in the house, because the limit of
     what immediately surrounds [it (e.g. air)] is in the house [in the
     primary sense (cf. 211a23- 26)]. He says, accordingly, that the now
     that is said to be so in its own right and primarily is indivisible and
     one in number, the same [now being] the limit of the past and the
     beginning of the future, and these are not different things; being
15   such, [the now] is not a part of time, since it inheres in every [stretch
     of] time, the same [now] being a beginning and an end, just like the
     point in every line, when [the line] is taken as one and continuous.
     For [the line] is everywhere divisible, and the division is at the
     point. Concerning this now he proved also in the [passage]
     concerning time [cf. 222a10-b7] that it is partless and indivisible
     and one and the same in every [stretch of] time, the limit of the past
20   and the beginning of the future. He proves these things here as well,
     assuming in advance as self-evident that the now is the extreme of
     the past, and that before this now – what he himself calls 'to this
     side of which', that is, to the past – 'there would be nothing of the
     future'. For before the end of the past how can there be something
     future? For it would not yet be future, but rather still past, existing
     before the limit of the past. Again, it is obvious that the now is the
25   beginning of the future, since nothing of the past is after it [i.e. the
     beginning of the future]. For how could something past be after the
     beginning of the future?[31] But Aristotle, having said 'to this side' of
     both [that is,] both the limit [of the past] and the beginning [of the
     future], made the argument unclear. But Eudemus [fr. 102 Wehrli]
     says, 'nothing to this side of the limit [of the past] nor to yonder side
     of the beginning [of the future]'. If it were proved, accordingly, that
30   these things, both the extreme [or last now] of the past and the first
     [now] of the future, between which there is no time either past or
     future, were the same and one in number, although seeming to be
956,1 two, [then] it will have been proved that this [now], which we call
     the same, simultaneously a beginning and a limit of time, is also
     indivisible.
         Having in this way, accordingly, accurately defined a 'now', with
5   which the argument is [here] concerned, he proves that the extreme
     of the past and the future (having called the beginning of the future
     an extreme as well, because a beginning too is a kind of limit and is
     partless) is one and the same. He proves it by inferring three
     impossible consequences of saying that the limit of the past and the
     beginning of the future are not one and the same now, but each a

---

[31] Simplicius 955,18-27 closely follows Themistius *in Phys.* 189,22-31.

different one. The first of these [consequences] is drawn from the [an argument by] division of this sort: If each is different, [then] either they are consecutive upon one another (here he takes 'consecutive' 10 [*ephexês*] instead [i.e. in the sense] of bordering [*ekhesthai*] and touching) or separated from one another, and there is nothing besides these [alternatives]. If, accordingly, it is neither possible for them to touch – not only because it is necessary that touching things have a limit as one [part], with which they touch, and what is limited as a different [part], but also because, as he himself says, it is impossible that something continuous be made of partless things...[32] Time is continuous, while its limits are partless: for here 15 this has been taken as agreed upon because they [i.e. the limit of the past and the beginning of the future] are hypothesized as limits;[33] for if they should not be partless, they could not be extremes, but rather other things would be the extremes of these. By syllogizing in this way, he will appear to be assuming what [was to be proved] in the beginning: for it was proposed to prove that the now is indivisible. For if it is proved, he says, that [the nows] are the same, it will simultaneously be clear that it is also indivisible.[34] But he uses this [latter premise] as self-evident when he says that it is 20 impossible for a continuous thing to be made of indivisibles. It is also possible to say that what is immediately being investigated is whether both these limits are one in number. This is the now, and in respect to this it has been taken as self- evident that two limits do not touch one another and make a continuous thing because touching things need to touch at parts [of themselves], but the limits are partless and indivisible each in its own right – since nothing [in 25 principle] prevents them from being separated and divided from one another, which indeed will [later] be proved to be impossible in the case of the limit [of the past] and the beginning [of the future]. It was proved in general that it is not possible for a continuous thing to arise out of partless things that touch each other. And through this [argument] it has been proved that the extremes are not two in the manner of things touching each other, while through the next [argument it is proved] that neither are they two in the manner of things separated; from which [arguments] it is inferred that [they 30 are] one and the same: it follows upon this that it is also indivisible, which is stated here not with respect to two limits, as [it was] earlier [cf. line 15], but with respect to the common [limit], with, indeed, the

---

[32] The sentence begins as though Simplicius intended to mention both alternatives (cf. *mête*), but breaks off after the mention of the first, i.e. that the nows touch; Simplicius takes up the second alternative, that they are separate, at 957,3ff.

[33] Reading *to perata* with MSS ACM, instead of *to ta perata* with Diels, F, and the Aldine.

[34] Reading *tauton, hama* instead of *tauton hama* (as suggested by Cecilia Trifogli).

fact that it is indivisible proved not from the [nature of the] continuous, as earlier, but from the fact that it is one. Thus, the argument does not assume what was [to be proved] in the beginning, since here it necessitates [only] that the two hypothesized limits be

957,1   partless, while there it demonstrates that what has been discovered to be one and the same [i.e. the now] is indivisible.

234a7-14: 'If each is separate, there will be time in between' to 'this will distinguish past from future time.'

[The remaining alternative is that the limits of the past and of the future might be separate, but then there would be time in between and time is divisible; hence the now would be divisible at one or another point. But then a part of the past (i.e. of what preceded the end of the 'now') would be future (because after the dividing point) and vice versa.]

5   Having proved that the extremes cannot be different in the manner of things touching each other, he proves next that neither will the extremes of time be two in the manner of things separated and standing apart from one another. For if they should be two [in this way], it is necessary that there be time in between. For in every continuous thing what is between limits is univocal [*sunônumon*, i.e. describable by the same name; cf. *Cat.* 1a6-11] with those [outer]

10   things of which they are the limits. If the limits that are in them [i.e. in lines] should be of lines, what is between [will be] a line;[35] for between all points there is line, and lines are univocal with one another, while if the limits should be lines, what is between the lines [will be] a surface; for lines are the limits of surfaces. Accordingly, what is between [any] two separate lines taken in the continuity of a surface is invariably a surface, and what is between two surfaces in

15   a [solid] body is body: in this case, even if we take surfaces of bodies that are separated, it is necessary that what is between be body, because of the non-existence of void. The same account [applies] as well in the case of motion. For between the limits of a motion, which they call moves, and which are in the continuity of a motion, it is necessary that there be motion. Similarly also in the case of time: for

20   this too is continuous, and nows are its limits. For this reason it is necessary that there be time between nows that are standing apart. For time has something exceptional [about it] in comparison with

---

[35] Translated according the the MSS; Diels obelizes. But *grammôn ta perata* answers to *hôn esti ta perata*; the idea is that the stretch between two limits, e.g. two points, in a continuous thing is the same kind of stuff as the outer two extents or magnitudes of which the two points or whatever are the limits.

the other continuous things, [namely] that all of it is continuous and
it is not detached from itself as line is from line, and surface from
surface when they are in different bodies. Accordingly, what is
between any limits of time is time. All time is divisible, because time
is continuous, and every continuous thing is divisible. If,   25
accordingly, this time should be divided, which is between the limit
of the past and the beginning of the future, and which he called the
'now' in the sense of the present, because it is between the past and
the future – if this [time], accordingly, should be divided, it will be
divided into a past and future [time]; for everything that is present
is so divided. With these things having been assumed in advance,   958,1
since this time in between, insofar as it is after the limit of the past,
is future, but insofar as it is before the beginning of the future, is
past, when the whole [time] between is taken as future there will
then be something of the past, which arises from the division, in the
future, and again when [the whole time] is taken as past, the future   5
which [arises] from the division will be in the past. It is possible to
infer the absurdity in respect to either part of the division, as
Aspasius did. For since the division of what is between produces a
past [part] and a future [part], if either of the segments of the
present is divided, it will again be divided into a past and a future,   10
and there will be something of the past also in the future and
[something] of the future in the past.

234a14-16: 'At the same time, the now would also not be [so] in
its own right, but by virtue of something else; for the division is
not in its own right'

[Further, the extended present would be present only by
containing a genuine point-like present].

I believe that this second argument is in refutation of some objection   15
that says that it is not at all absurd that there be a now, in the sense
of the present, between the past and the future. [Aristotle] says that
this in-between is not a now in its own right. For [the now] in its own
right is partless, while this [one] is divisible, because it is [a stretch
of] time. Even if it is called 'now', accordingly, it is called so not
because the name belongs to it in its own right, but because [it does   20
so] by virtue of something else, [namely] because the now in its own
right [i.e. the present instant or temporal limit] is in it. That this
now is not [the now] in its own right, such as that which is in the
strict sense between the past and the future must be, is obvious
from its being divided. 'For the division is not' of the now 'in its own
right', but rather of the [now] that is [so] called in a broad [sense]   25

and by virtue of something else. There is also another reading as follows: 'For the division is in its own right.' He would be saying [on this reading] that this in-between is a now by virtue of something else; for the [point of] division is the now in its own right, but what is divided is not.

959,1   **234a16-20**: 'Besides this, some of the now is past' to 'it is necessary that the now in each [i.e. past and future] be the same.'

[And if the now were divisible, part of it would be past and part future, and not always the same part; nor would the now be a single thing.]

Having proved previously that for those who say that the limit of the past and the beginning of the future are not the same and one in
5   number, but have time in between, it follows that some of the future time will be in the past and [some] of the past in the future, he proves here that of this now itself which is hypothesized to be between the past and the future, there will be a past [part] and a future [part]. For if this time between the limits, which appears to
10   be now, since it is neither past nor future, should be divided, it is obvious that its parts will also be now. For the whole [time] was [said to be] now. But every division of time produces a past and a future, distinguishing each, so that some part of the now, although it itself is now, will be partly past on the one hand, future on the other. But it is absurd to say that some of the present now is past and some is future. That some of it is past and some future follows,
15   in fact, from defining the now in a broad [sense], since the now in the strict sense, which is partless, distinguishes the past and the future. If this in-between [now] should be cut again at another [point] and not at [the point] at which it was formerly cut, but either near the beginning of the future or near the limit of the past, if [it is cut near] the beginning of the future, it will then, while in the future, be again
20   in the past, and if [it is cut near] the limit of the past, it will then, while in the past, since it is a part of it, be again in the future;[36] so that[37] the same part of the now will in turn be past or future. Rather, the same [part] migrates, as each division produces a past

---

[36] Inserting *palin en tôi mellonti* after *autou*, as suggested by David Sedley (I originally inserted it after *pareléluthoti*, which also gives satisfactory sense); Diels wrongly inserts *mellonti palin en tôi* before *pareléluthoti*.

[37] Reading *hôste*, as suggested by David Sedley, instead of the Aldine supplement *kai houtô*, which is adopted by Diels. As Sedley observes, the scribe's eye may have skipped from one *palin* to the next, thus causing the lacuna.

[part] and future [part] of what is divided. Having proved that 'the same thing will not always be past' and 'future', he adds that 25 'neither will the same thing' always be 'now'.[38] The reason is that this in-between [time], since it is time,[39] is divisible into forever divisibles, and all the cuts in it will be into nows that are present, since both this whole time, in which the cuts [are made], is now, and also the divisions, each becoming now at a different time. If, accordingly, the in-between time is now, the result will be that the 960,1 things that occur in the same now are not contemporaneous, on account of the many [nows] in a broad now. If, accordingly, the extremes are neither two in such a way as to touch one another nor two in such a way as to be separated, it is necessary that the limit of the past and the beginning of the future be one and the same now. 5

234a20-24: 'But if indeed [it is] the same, it is evident that it is also indivisible' to 'it is obvious from what has been said.'

[If the same now is the limit of the past and of the future, it is indivisible, for the reasons given above.]

It follows upon its having been demonstrated that the limit of the past and the beginning of the future are one and the same, he says, that it is also indivisible. For if it were divisible, the 10 above-mentioned absurdities would again attend the division of it. For there will be a past and a future of the now, and the same thing will not always be past nor the same thing always future, but rather not even the now [will be] always the same. That it is indivisible is obvious also from the fact that it is the limit of the past and the beginning of the future, and that a limit and a beginning are 15 indivisibles. But he took this as agreed upon when he said, 'because a continuous thing is not made of partless things' [234a7], while here he proved, not that the limit and the beginning are each indivisible, but rather that both are one and the same, and one in the manner of an indivisible (for a day is also one thing, but not in the manner of an indivisible).[40] Both these nows were discovered to be one and the same, according to what was hypothesized, and one in the manner of indivisibles. 20

---

[38] The manuscripts of Aristotle's text have *oude dê to nun to auto*, i.e. 'neither, indeed, [will] the now be the same'.

[39] Omitting Diels' supplement *nun*; David Sedley suggests reading *ho metaxu houtos khronos, <khronos> ôn*, which improves the syntax.

[40] Closing the parenthesis after *adiaireton*, instead of after *hêmera* with Diels.

234a24-34: 'That nothing moves in a now' to 'it is obvious that neither is it at rest.'

[Nothing moves in a now; for if it did, it could move faster and slower. Suppose the faster thing moves over the line AB in the now N; the slower will move over the lesser line AC. Thus, the faster thing will traverse AC in a time less than N; but N was posited to be indivisible. Again, nothing is at rest in a now; for only what is so constituted as (i.e. by nature able) to move under given circumstances can be said to be at rest under those circumstances.]

Having proved that the now, taken as simultaneously a beginning and a limit, is one and the same and indivisible, he draws a still
25   more paradoxical [consequence], that nothing either moves or is at rest in it. He proves it through several arguments, of which the first is as follows: If something moves in the now, the now will be divisible; but this, in fact, is impossible (for it has been proved to be indivisible); consequently, a thing does not move in the now. The additional premise is evident, [namely] that it is impossible for the
961,1 now to have been divided. He proves the conditional premise by assuming in advance what has already been agreed upon, that it is possible for something to move faster too in any [stretch of time] in which it is possible for it to move slower. Through these [premises] he proves by way of an illustration that if something should move in the now, it is necessary that the now be divided. For let N be the now
5   in which something is said to move, and let the faster-moving thing move [over] the [line] AB in N. 'Therefore the slower thing will in the same now move [over] a lesser [line] than AB, for example, the [line] AC. Since the slower thing has moved [over] the [line] AC in the now, the faster will move [over] the [line] AC in less than the now' [quoted from Aristotle, with small variations]. Consequently, the now is divisible. Next he concisely proves that neither is anything at
10   rest in the now, using the same form of demonstration. For if a thing does not move in the now, neither is it at rest; but, in fact, a thing does not move [in the now], as has been proved; consequently, neither is it at rest. Again, the additional premise is clear, since it has been proved immediately before, but he suggests the conditional premise from the definition of being at rest: 'For we call at rest' not every motionless thing, but rather 'that which is so constituted as to
15   move, but is not moving when it is so constituted' as to move (therefore we do not say that a new-born puppy is at rest in respect to seeing, because it is not then so constituted as to move [i.e. act] in respect to it) and where it is so constituted (for water-creatures are not at rest when they are not moving on land) and in the way in

which it is so constituted (for swimming things are not at rest when they are not walking); for being at rest was defined in this way. If, accordingly, 'nothing is so constituted as to move in the now, it is obvious that neither is [anything] at rest'.

234a34-b5: 'Further, if the now is the same' to 'for the now is   20
the same extreme of both [i.e. past and future] times.'

[If the same now is the limit of the past and of the future, and if
a thing can move in the whole past and be at rest in the whole
future (or vice versa), then it will move and be at rest in the
same now, if it is once allowed to do either in a now.]

A second argument proving that it is not possible for something to move or be at rest in the now in the strict sense: he proves this too according to the second form of hypothetical [syllogisms], by reducing to the impossibility that the same thing is simultaneously   25
moving and at rest. For taking what has previously been proved, [namely] that the now, being one and the same in number, is the limit of the past and the beginning of the future, he says that if something is taken as moving during the whole of past time, it will also be moving in any [bit] of it whatever among those at which it is   962,1
so constituted as to move. If, accordingly, it is so constituted as to move also in the now that is the limit of it [i.e. of past time], as the argument hypothesizes, it is obvious that it will move in it as well. If the same thing is taken as being at rest during the whole of future time (for it is possible so to hypothesize), it will also be at rest in any [bit] of it in which it is so constituted as to be at rest. If, accordingly,   5
it is so constituted as to be at rest in the now at the beginning of the future, the thing that is at rest in the whole future will be at rest also in it. Consequently, in the same now, which is the limit of the past and the beginning of the future, something will simultaneously move and be at rest in the same respect [*kata tauton*].[41] For the now is indivisible; therefore it will not move in this [part] of it, and be at   10
rest in that [part]. He has added the [phrase] 'at which it is so constituted as to move', because moving things are so constituted as to move in portions of time, but not so constituted [as to move] in its limits. If, nevertheless, someone were to hypothesize that they move in these as well, the absurdity that has been proved will follow.

[41] There seems to be a reminiscence here of Plato *Republic* 436C5-6; 'Is it possible for the same thing simultaneously to move and stand still in the same respect [*kata to auto*]', where, however, the context reveals that *kata to auto* means specifically 'in respect to the same part of itself'; the idea, here as there, is to exclude paradoxes involving rotation in the same place.

234b5-9: 'Further, we call being at rest' to 'consequently, it is
15   necessary that a moving thing move and a thing at rest be at
rest in time.'

[A thing is said to be at rest if it and its parts are in one and the
same state now and earlier; but there is no earlier in a now;
hence a thing cannot be at rest in it. Thus a thing moves or is at
rest only in a divisible stretch of time.]

He infers a third [time] that it is not possible to be at rest in the now.
The syllogism would be categorical in the second figure, as follows:
to be at rest is for a thing itself and its parts to be alike both now and
before. In the now, since it is indivisible, there is no now and before.
20   Consequently, to be at rest is not possible in the now. Having
demonstrated so manifestly what was proposed [for proof], he
reasonably added that 'it is necessary that both a moving thing
move and a thing at rest be at rest in time', since it is not possible
either to move or be at rest in the now.

234b10-20: 'It is necessary that everything that is changing' to
'it is evident, accordingly, that everything that is changing will
be divisible.'

[What is changing is divisible. For part of it must be in its new
state or position and part in the old. That to which a thing is
changing may be the first stage of a change, e.g. grey in the
change from white to black.]

25   Having proved earlier that the magnitude on which a motion
[occurs] is divisible into forever divisibles, and also the motion itself
and the time, he proves here that it is also impossible for the moving
thing itself to be without magnitude and indivisible. What is said
here has been added in consequence of what was said immediately
before. For having proved that nothing either moves or is at rest in a
30   partless now of time, while both moving things move and things at
rest are at rest in divisible time, he proves here that not only is the
963,1   time in which the motion [occurs] divisible, but also the moving
thing itself and the motion and the moving and the magnitude on
which the motion [occurs]. All are alike, because if one is divided the
others will also be divided. Meanwhile he proves here that the
moving thing is divisible, assuming in advance as agreed upon that
every change is from something to something, and constructing the
5   argument on a [logical] division as follows, he says: Since a changing
thing changes from something to something, it is necessary that in

the [process of] changing both it itself and its parts either still be in that from which it is changing, or in the thing to which it is changing, or in both or in neither, itself (once again) and its parts, or [finally] some of it in that from which it is changing, and some in 10 that to which it is changing. Besides these, there does not appear to be any other segment to the [logical] division. Accordingly, if it should be proved that none of the other segments is sound, and if there should be left the one that says that some of it is in that from which it is changing, while some is in that to which it is changing, what is changing will, obviously, be revealed to be divisible. But it is obvious that when it is in that from which it is changing, it is not changing (for then it is at rest and is not yet changing), nor again 15 when it is in that to which it is changing (for then it will have changed and not be changing), nor is it possible that the whole of it be in both, [i.e.] that from which it is changing and that into which it is changing (for it will both be apart from itself and at the same time will have changed apart from changing), nor will it be such that both itself and its parts will be in neither. 'In neither' is said in the sense 20 of 'nor in what is between', as will be obvious; and this is the most impossible of all. There is left, consequently, that some of the changing thing is in that from which the change [is taking place], and some in that to which it is changing; and if this [is the case], it will be divisible. Not that when [something] is changing from white to black, it will be [partly] in the white and [partly] in the black: for 25 it is necessary that it first come to be in the grey. That is why he added the [sentence], 'I mean what it is changing into first along the change, for example from white, grey – not black.' Therefore it is not necessary that the changing thing be either in that from which or in that to which, but [only] in the first thing along the change; the in-between is a first [stage] of that to which, as to an end, and grey is [a first stage] of black. For [the changing thing] travels through this 30 [i.e. grey] away from white toward black, so that when he says that it is able not to be[42] in that to which it is changing, one must 964,1 understand not that to which it changes first, but rather that toward which it is heading as toward an end. Similarly too, when he says, 'nor in neither', he says that not as though it could be in between, but rather as if he said, 'neither in the white nor in the black nor in any of the things in between'. 'Accordingly it is evident,' he says, 5 'that everything that changes will be divisible', but none of the parts changes in its own right; rather, as many of these that do change, change incidentally. For when a body moves, the things in it also move, for example the plane, the line, and the point.

[42] Aristotle says, 'it is not necessary that it be'; Simplicius' *ouk hoion te einai* has been translated accordingly, rather than as 'it is not able to be' (which would be the more natural sense of the Greek).

From this argument it is possible to solve the argument that denies motion from the [premises] that a moving thing must be either in that from which the motion [occurs] or in that to which the motion [occurs], and that neither is possible; for while it is in that from which it is not yet moving, and while it is in that to which it is no longer moving, but rather at rest. For it was proved that as a whole it is in neither, but some of it is in this and some in that. In [his commentary on] this [passage] Alexander, dragging everything into his own hypothesis about the soul which says that the soul is inseparable from the body, says that[43] 'this is obvious from what is said here as well, since the soul is incorporeal and partless, and a partless thing does not move, if it is necessary that some of a moving thing [be in that] from which it is moving, and some to which it is moving, but what does not move is not separated. In avoiding this absurdity,' he says, 'some people hang some body around it [i.e. the soul] as a vehicle,[44] and do not notice that, by this, they are either saying that body passes through body, if, at all events, the soul penetrates the entire body and is [itself connected] with body [as its vehicle], or that they are separating it [i.e. the soul] from this [i.e. the vehicular body] and moving [the soul] in its own right in its penetration into bodies.' With respect to [the argument that] the soul does not move because it is partless, it must be said, I believe, that it does not move [with] a corporeal motion, but that in deliberating and intending and discriminating and speculating it nevertheless moves [with] motions that pertain to the soul. I think that the propounded definition of motion could fit motions of the soul as well, or at all events most of them, such as have something potential[45] in their moving. Even if it does not fit but is separate, one must specify them [i.e. the motions of the soul] in accord with some other type of motion. This is maintained also by Theophrastus, the leading associate of Aristotle, in the first of his [books] *On Motion* [fr. 53 Wimmer], where he says that 'desires and appetites and feelings of anger are corporeal motions and take their beginning from this [i.e. the body], but as for those [motions] that are discriminations and speculations, it is not possible to refer these to something else, but rather their beginning and activity and end are

---

[43] Diels begins the quotation from Alexander with the following sentence, but diction and style (e.g. omission of *en tôi* before *ex hou* and *eis ho*) suggest that Alexander's words begin here.

[44] Cf. Proclus *Elements of Theology* proposition 205, with Appendix II in the edition by E.R. Dodds.

[45] Reading *ti dunamei* with MSS AF; Diels emends to *to dunamei*. Aristotle defines motion as 'the actualization (*entelekheia*) of what is (such-and-such) potentially (*dunamei*)' (201a10-11), e.g. transformation is the motion of what is transformable; a motion of the soul would be the actualization of whatever is in it potentially.

in the soul itself,[46] if indeed the mind is something better and more
divine [than the body], since it enters from without and is
completely perfect.'[47] And to this he adds: 'Concerning these things, 5
accordingly, one must consider whether [the mind] has some
separateness with respect to its definition,[48] since it is agreed that
these [i.e. discriminations and speculations] too are motions.' Strato
of Lampsacus, who became a disciple of Theophrastus', and is
counted among the best of the Peripatetics, also agrees that not only
the irrational soul but also the rational [soul] moves, when he says
that the activities of the soul too are motions. He says, accordingly, 10
in his [book] *On Motion* [fr. 74 Wehrli], besides many other things,
this as well: 'For someone who is thinking is always moving just as
someone who is seeing and hearing and smelling; for thinking is the
activity of the intellect [*dianoia*] in the way that seeing is of sight.'
Also before this statement he has written: '...since, accordingly,
most of the motions are responsible [sc. for thinking], both those
[with] which the soul moves in its own right when it is considering 15
and those [with] which it previously moved by [the action of other]
motions [i.e. those of the senses].[49] And it is obvious: for whatever it
[i.e. the soul] has not previously seen, it cannot think, for example
places or harbors or pictures or statues or people or any of the other
things of this sort.' Now, from this it is obvious that the soul moves
according to the best of the Peripatetics, even if it is not [with] a
corporeal motion. Aristotle himself in the last book of this treatise 20
[254a29] says: 'For imagination and opinion are thought to be kinds
of motions.'

With respect to the other argument [based] on the vehicle, it must
be said that it is not because the soul is separate from this body that

[46] Cf. Aristotle *de Anima* 408b15-19 on sensation as a motion that comes from
outside the soul, and recollection as a motion starting from the soul; Aristotle then
describes *nous* as something imperishable and separate from us.

[47] The phrase 'if...perfect' seems to be the beginning of a new sentence, if the MSS
reading *ei de dê* is retained. Diels' conjecture of *ge* for *de* ('if indeed, at all events')
gives satisfactory sense with minimal alteration of the text.
Aristotle says (*de Generatione Animalium* 736b27-28) that the mind (*nous*), unlike
the nutritive and sensitive souls or faculties, is divine, pre-existing the body and
entering it from without. Alexander took 'mind' not as the entire human intellect, but
as the so-called 'active intellect' (*de Anima* 430a15-16) which, in his view, was not
part of the human soul at all but rather God resident within us so long as we are
alive.

[48] If the soul has motions that are independent of the body, i.e. motions that, unlike
desires, appetites, and anger, originate and end in the mind, then the definition of
mind will include no reference to the body, and thus be separate; cf. *de Anima*
403a3-b9, with 408b18-19.

[49] Reading *aitiai*, 'responsible', with Diels and the MSS (Wehrli, following
Poppelreuter, emends to *hai autai*, 'the same'), and *hupo tôn kinêseôn*, again with
Diels and the MSS (Wehrli, following the Aldine, emends to *hupo tôn aisthêseôn*, 'by
[the action of] the senses').

we say that that vehicle is attached to it [i.e. the soul]; for even when
[the soul] is in it [i.e. the body] it is separate in respect to the essence
25 [*ousia*] of it [i.e. the body]. For if [the soul] has activities that are
separable from it [i.e. the body], all the more does it have its essence
separate, as Aristotle put it in the [treatise] *On the Soul* [430a17].
The soul is not, indeed, said to be in this body locally, as in a bucket,
but rather relatively [*skhetikôs*], so that to be separate from it, it has
no need of some other body that is moving locally. For the
unsuitability of the body suffices for it not to receive the
30 illumination of the soul: this is what the soul's being separate is. But
966,1 they demonstrate that that [i.e. the vaporous] vehicle is attached to
the soul because, since [the soul] is in the world [*enkosmios*] and
governs in different parts of the world [*kosmos*], it invariably has a
vehicle proper to that part in which it governs, which [vehicle] is
animated by it. When, accordingly, [the soul] governs in the air, the
thing that is fastened to it is vaporous [*pneumatikon*], just as here
5 [i.e. in the body] this is the oyster-like thing [i.e. the visible corporeal
vehicle of the vegetative soul].[50] If this is the reason for the different
vehicles, it is not necessary, perhaps, that when [the soul] is in this
[i.e. the body] it also have the vaporous [vehicle], but if it should also
have it, it is not necessary that that [i.e. the vaporous vehicle] be in
this [i.e. the body] in such a way that body passes through body,
while I do not even think that this is absurd, [namely] that higher
and finer things, not of like nature, should pass through crasser and
10 more material things;[51] for in fact solar light, according to those who
say it is body,[52] is said to pass through the whole air, and the
heavenly spheres, which are full and not just [hollow] vaults, since
they are continuous with respect to the centre, pass through the
whole [spheres] that are next [inside them]. Let these things,
accordingly, which, as Alexander himself also said, are extraneous,
find this sort of resolution on my part.
15   They pose a dilemma nicely in regard to what was said by
Aristotle, [namely] that a moving thing has something in that from
which it moves, and something in that to which [it moves], or in
between; for if this is true, how did he himself, in the first [book] of
this treatise [186a15], when he was impugning Melissus for
speaking of the beginning of a transformation, add: 'as though the

[50] Cf. Olypiodorus *Comm. in Plat. Phaed.* Norvin 143.14, 240.13; *Comm. in Alc.*
Westerink I 5.9; Proclus *Comm. in Timaeum* Diehl III 237.26, 29; the idea seems to
derive ultimately from the humorous analogy in Plato *Phaedrus* 250C4-6, where,
however, the point is that the soul is contained by the body like an oyster in its shell.
[51] Cf. 531,3-9; 616,23-617,2; 623,32-624,2; 643,18-26, on the interpenetrative
power of the celestial body.
[52] E.g. Stoics like Chrysippus, criticized by Alexander in *de Mixtione* 218,8-9,
Mantissa 138,4-139,28; so too Simplicius *in Cael.* 12,28; 16,21; 130,31-131,1, but cf.
in DA 134,13-20.

change did not occur all at once [*athroos*]'? For he speaks as though   20
an entire thing could be transformed all at once and not portion by
portion. For in fact milk and water that are frozen and that are
warmed seem to change all at once as a whole, when they are
transformed, and not part by part. If a transformation occurs in this
way, even if the moving thing were a body, it would not be necessary
for some of the changing thing to be in that from which the change
[is occurring], and some in that into which it first changes. For what   25
changes all at once is wholly in the other thing into which it has
changed. Such being the puzzle, Alexander says that this puzzle
does not address what is said here: for here the argument concerns
what is changing, that is, what has not yet changed but rather is
still changing, in the case of which there is every necessity that
some of it be in that from which, and some in that to which the
change first [occurs]. But if at all, the puzzle would be relevant to   30
[the argument]: If not everything that has changed has changed
through changing, it would have changed without previously
changing. The puzzle is posed, accordingly, not with respect to what   967,1
is changing, but with respect to what has changed, if it is possible for
something to have changed without previously changing. For in fact
Aristotle, in what was said earlier [232a10], reduced this argument
to absurdity, when he said: 'so that there will be something that has
walked without ever walking.'

Some say that it is because of this puzzle that [Aristotle] has
produced the proof here in the case of something moving locally; for   5
in this case it seems sound that some of a moving thing ought to be
in that from which, and some in the thing to which the change first
[occurs]; and this is confirmed by the fact that [Aristotle] took that
which moves over a magnitude; for this moves locally. Others say
that [Aristotle] proves it for everything that moves [*kinoumenon*],
and not only for what moves, but also for everything that changes   10
[*metaballon*].[53] For he says that 'it is necessary that everything that
changes be divisible'. The argument is not, consequently, only about
something moving locally. Further, the [argument] by which he
proves that what changes is divisible pertains not only to things

[53] At *Metaphysica* 1067b15-1068a10, Aristotle explains that there are three kinds
of motion, namely, local, qualitative (i.e. transformation), and quantitative (i.e.
increase and decrease), all of which are changes between contraries (e.g. near far,
black white, small large). However, change between contradictories (e.g. coming to be
and perishing) is not a motion. At 200b32-201a3, Aristotle seems to use motion
loosely of all four kinds of change. Either way, it is essential to recognize that *kinêsis*
and related verbal forms may have a much wider sense than the English 'motion' or
'move', often approaching the idea of 'process', i.e. that which goes on in changes such
as that from black to white. I have kept the traditional rendering 'motion' for *kinêsis*
in part because 'process' has no corresponding verbal form in ordinary English
('proceed' has the wrong sense), in part because 'motion' usually causes no difficulty.
Where intelligibility requires, the wider sense of *kinêsis* will be indicated.

moving locally, but likewise to all things that change. For every
change is from something to something, as he himself said. If,
15   accordingly, what changes is divisible because of this, [namely]
because change occurs from something to something, and 'from
something to something' embraces not only things moving locally
but rather all changing things together, [then] either all changing
things that share in a whence and whither will be alike divisible, or
none. That his argument is not about locally moving things alone,
but rather about all things [that change], he indicated also through
20   the example that he cited: for he took white, grey and black, in
respect to which transformation, and not local motion, is
accomplished. It was with a view to transformation that he said in
the first [book] of this treatise, 'as though the change did not occur
all at once'. But he could have been using white, grey and black as
25   an example here not because he was speaking about transformation,
but for the sake of proving what the [immediately] next thing is to
which a change occurs. However, in the next [arguments], he uses in
the case of all moving things the fact that it has been proved that
they are divisible in general. At the end of the book [241a28] he says
even about things that change according to contradictory pairs –
these are things that come into being and perish – that a part of
30   these things too is in each portion of the contradictory pair, as
though holding it to have been proved that not only what moves, but
also everything that changes, is divisible.

It is worth remarking that in the case of things moving locally it is
necessary that one [part] be in that from which, and another [part]
968,1   in that to which. For in the case of these, the 'whence whither' is this
way, but in the case of transformation and increase and decrease it
does not seem necessary that it be this way, but rather in the case of
these it is possible both for a transformation to occur all at once and
[for a thing] to increase all at once. For the addition seems to occur
simultaneously to every portion of a thing that is increasing, and
5   similarly decrease too is all at once. Alexander says: 'There would be
motion [i.e. a process of change] also in the case of things that
change all at once by virtue of the fact that for these too there is
invariably something between, along which the change occurs, and
the changing thing cannot change in [one and] the same respect [i.e.
increase, decrease, or the like][54] from this to that except through
what is between. For the magnitude into which the increasing thing
10   changes, and everything that is added to it so that it may increase,
are divisible; and likewise too what departs so that it may decrease.
There are also things in between the qualities into which change in

---

[54] Reading *kata tauton* with MSS CM, rather than *kat'auta* with MS A and the
Aldine, followed by Diels, who explains in the apparatus criticus that *auta* = *auxêsin
kai meiôsin*, i.e. 'according to increase and decrease'; cf. 962,8-9.

respect to transformation occurs, through which the thing being transformed changes toward being of such a kind, even if it changes simultaneously as a whole. For, whatever it may change into, there will be some first [stage] of that, which is between that from which it was changing and that into which it has changed. This is also why 15 every motion is in time. It is necessary,' Alexander says, 'to inquire further concerning this. For having proved shortly afterward that quality too, according to which transformation occurs, is divisible [235a18], he says that [it is so] "incidentally". For it is co-divided [i.e. divided in conjunction] with the division of the thing that is transformed, since this too changes portion by portion. But,' he says, '[we may reply that] it is better to understand the [expression] "as though the change did not occur all at once" not as though it is said 20 of the whole moving thing, but rather as if a part of a moving thing changes all at once, while the whole thing does not yet [change]. For a part of freezing milk changes all at once, but not the whole thing, a sign of which is that freezing does not always occur in a like manner in the case of greater and lesser [amounts]. For in every case, what is more easily affected either by nature or by virtue of being closer to 25 what produces [the change] changes more quickly. A part of something that is being sunburnt too changes all at once. For the surface that is turned toward the sun [changes] first; this [surface] was not the moving thing [i.e. the thing undergoing a process of change], but rather that of which this was [the surface]. Even if all that which is increasing increases, nevertheless it does not [do so] simultaneously; for some portions are closer than others to the provisions that supply them their increase.' Alexander, then, 30 contributed these things to this passage.

The clever Themistius does not accept from Alexander the [statement] that even in things that seem to change all at once, as in the case of freezing milk and a darkening body, one part changes 969,1 earlier, and another later. For he says [192,8-9] that perception proves that there occur some all-at-once changes and transform- ations of bodies. But [we may reply that] evidence from perception in these things is not safe. For small portions both of milk and of a 5 darkening body that are not tra ᴉformed simultaneously with the rest might escape perception; but one must rather say that if indeed there is any part at all, even a short one, that is transformed all at once, which indeed Alexander grants, since that too is partitionable, [then] if it is transformed as a whole simultaneously, what is said by Aristotle here is not true, [namely] that everything that changes has 10 one of its parts in that from which it is changing and another in that to which it is changing. But if not even the part is transformed as a whole simultaneously, what was said against Melissus in the first book, [i.e.] 'as though the change did not occur all at once', would be

misleading [because it implies that there is change all at once], so
that the puzzle still remains. Themistius, however, says [192,12-22]:
'What has occurred to us to help out the argument is this: we think
15   that Aristotle does not believe that he needs in addition an
argument concerning things that change all at once [proving] that
they are divisible. For if we say that those things change all at once,
all the parts of which simultaneously and at one time are
transformed or increase, it is obvious that it is in the case of things
that possess some parts that there is change of this sort. As for
things that possess no part, how would they be transformed
20   simultaneously at all of their parts? Therefore, either changing all
at once is not this, [namely] changing at all portions together, or, if
[changing] all at once is this, it is ridiculous to inquire whether such
things are divisible. Therefore Aristotle did not even produce his
argument at all for the case of these things, but for the case of things
that change according to the other forms [of change].'

25   **234b21-29**: 'Motion is divisible in two ways' to 'so that the
whole motion is the motion of the whole magnitude.'

[Motion is divisible both according to the time taken, and
according to the motions of the parts of the moving thing. Let
AC be a moving magnitude, and its parts AB and BC move
with the respective motions DE and EF; it is the combined
motion DF that will be the motion of AC as a whole. Proof: AC
cannot move with the motion of something else (i.e. with the
motions DE or EF, which are those of its parts).]

Having proved that every moving thing is divisible, he proves here
that motion too is divisible and divisible in two ways, in 'one way by
time', because every motion is in time, and every time is divisible. It
30   has also been proved that a thing that is moving uniformly moves
970,1 less in less time and more in more [time]. This thing [i.e. the motion]
that has a more and a less is divisible in respect to time. For in fact
that [i.e. time] has a more and a less because it is divisible. But he
both has already said many things about the division of motion in
respect to time, and he will shortly afterwards speak again [of it].
5   Having proved immediately before that what moves is divisible, he
next proves here the other way of division in respect to motion,
according to the division that occurs of the moving thing. For since
the moving thing, being divisible, has a certain interval and length,
it is necessary that its motion, which is in the whole interval of the
moving thing, itself too have the same interval as the moving thing.
10   If, therefore, the extent of the motion is equal to the extent of the

moving thing, which is divided into parts, it too would be divided in the same way as the parts of the moving thing, so that the whole argument is as follows: if the motion of the whole is the same as that of the parts when added together, the motion would be divided according to the magnitude that is moving. Since the conditional premise is self-evident, if the antecedent is true, the consequent will also be true. That the antecedent is true he proves through the 15 following illustration: let AC be a moving magnitude, and AB and BC be its parts, and let DE be the motion of AB, and EF [the motion] of BC. For if the whole is moving, it is necessary that the parts move too. I say, accordingly, that DF will be the whole motion of the whole magnitude AC as well. He demonstrates this through many [arguments], and once it is demonstrated, it is obvious that the 20 motion would be divided according to the magnitude. He proves it [i.e., that DF is the motion of AC] according to the first argument by taking it as something self-evident that none of the things that move moves [with] the motion of another thing. This is obvious to those who reflect upon the types [*eidos*] and parts of motion. For something moving locally will not move [with] the motion of something that is being transformed, nor [will] that [which is 25 moving with] the [motion] of AB [move with] the [motion] of BC, but rather each [will move with] its own [motion]. If, accordingly, AC moves as a whole, it is obvious that its parts, AB and AC, are also moving. For thus it will be a whole moving thing. And DE will be the motion of AB, and EF [the motion] of BC, since these are the parts of the whole motion DF. For even if the parts of a continuous thing do not move in their own right in the motion of the whole, they nevertheless do, at all events, move in the whole. By virtue of this 30 the motion too is divisible, but it has not been divided; for it would have been divided if each of the parts of the whole were moving in its own right. If, accordingly, the [motion] DE is that of AB, and EF that of BC, it is obvious that it is necessary that the whole [motion] DF 971,1 also be that of the whole [moving magnitude] AC. For it is not possible that it be the [motion] of another thing, by virtue of the supposition, in general, that nothing moves according to the motion of another thing. If this is supposed, the whole motion DF will be neither that of any one of the parts of the magnitude AC nor of any other thing besides these [parts]. Thus, accordingly, what was 5 proposed has been proved in one way: this was that the motion of the parts added together will be the whole motion of the whole magnitude made of the parts. The converse, accordingly, is also true, that if the whole [motion] is that of the whole, its parts are those of the parts, upon which it follows of necessity that the motion is divided according to the moving magnitude. But why did he not take some motion DF of the magnitude AC, and proving thus that

10   the [motion] DE is that of AB, while the [motion] EF is that of BC,
infer that the motion is divided according to the magnitude? [We
may reply] that[55] this would be to assume what was [to be proved] in
the beginning, and not to demonstrate what has been proposed [for
proof] through other [premises]. And in fact if the whole [motion] DF
had been taken from the beginning as that of the whole [magnitude]
AC, it would not invariably follow that its parts too were those of the
parts [of the magnitude]. For what if the motion was indivisible? If,
however, the parts are the [motions] of the parts, it is necessary too
15   that the whole [motion] be that of the whole [magnitude]. For it is
possible for a whole thing to be both indivisible and divisible.

234b29-34: 'Further, if every motion' to 'the whole motion
would also be that of the magnitude ABC.'

[Further proof: if every motion is the motion of something and
of that thing only, and if DF is not the motion of any of the
parts of the magnitude AC (for they have their own motions),
then it must be the motion of ABC as a whole.]

This is a second argument, proving that the whole motion DEF is
20   that of the whole magnitude ABC. It is proved on the [basis of the]
assumption in advance that the [motion] DE is that of the
magnitude AB, and the [motion] EF that of BC, and on [the basis of]
the axiom that says that every motion is [the motion] of some
moving thing. The proof proceeds from [an argument by] division of
this sort: the whole motion DF is that of either one of the parts of the
magnitude AC, or of the whole AC, or of some other thing besides
25   these. The division is exhaustive. If, accordingly, the whole [motion]
is not that of either one of the parts (for each [of the assumed
motions] is that of a part, the [motion] DE that of AB, and EF that of
BC: so that the whole [motion] is that of neither of them) – but in
fact neither will the motion DF be that of any other thing. For if it
were wholly [the motion] of some other thing, its portions would also
be those of the portions of that thing (for [that thing] would not be
moving if it were partless; this he takes as self-evident and as
972,1 already having been proved by him in the argument preceding this
one). But in fact the parts DE and EF of the [motion] DF are those of
the parts of the [magnitude] AC, [namely] AB and BC, and these
parts of the motion cannot be [the motions] of other parts as well [i.e.
of the parts of some other thing]: for it is not possible that the same
and numerically one [motion] be that of many things. What is left,

[55] Correcting the misprint *oti* to *hoti*.

consequently, is that the whole motion DF is that of the whole 5
magnitude AC.

234b34-235a10: 'Further, if there is another motion of the
whole' to 'and it is necessary that it be [the division] of
everything partitionable.'

[Further proof: if there were another motion of the whole
magnitude AC, e.g. HI, we could subtract from it the motions
DE and EF of the parts of AC, in which case either HI = DE +
EF = DF, or there is a remainder KI; but KI will not be the
motion of anything, which is impossible; and similarly if HI <
DE + EF. Thus HI = DF.]

Having proved that the whole motion DF is that of the whole
magnitude AC by virtue of the fact that it cannot be the [motion] of
any other magnitude, he proves the same thing here by virtue of the
fact that no other motion besides DF can be that of the magnitude 10
AC. For if it is possible, let some other [motion] besides DF, [say] HI,
be that of the whole magnitude AC. The parts, then, of the [motion]
HI will be [the motions] of the parts of AC. And if the parts of the
motion HI, for example HK and KI, are equal to the parts of the
motion DF, [namely] DE and EF, it is necessary that they be also the
same, and that the whole [motion] HI be equal to and the same as
the whole [motion] DF. For of each [magnitude] there is one motion. 15
And it is not possible that there be any other motions either of the
parts of AC along with the [motions] DE and EF or of the whole
along with the [motion] DF. But if the motion HI should be divided
into the motions of the parts [of AC], it will be the same as the
[motion] DF: for it is divided into the parts of that [motion]. But
[Aristotle] himself, instead of saying 'it will be the same as the
[motion] DF', since it is divided into the parts of that [motion], said
'the [motion] HI will be will be equal to the [motion] DF'. If, when 20
[parts] equal to the [parts] DE and EF are subtracted from the
[motion] HI, some other part should be left over, for example KI, this
part of the motion HI 'will be the motion of no [part]' of AC: 'neither
of the whole' (for the whole [motion] HI was that of the whole) nor of
any part of it; for it does not have any other part besides AB and BC,
and the motion HI was divided into two [parts] equal to the
[motions] DE and EF of these. Having said 'nor of any other' of the 25
parts, he neglected to add, since it is clear, 'because it has no other
part'. That it [i.e. HI] is not [the motion] of any of the [parts] into
which it was divided, he proved by saying 'because there is one
[motion] of one thing'. For the same part [i.e. of a motion] cannot

simultaneously be that of some one of the parts DE and EF and that
of the whole, but neither will it be that of some other [part] from
outside, too. As the reason for this, he added the [fact of being
973,1   continuous]:[56] for the motion HI is continuous, and a continuous
[motion] is that of something continuous and one. There is no other
portion in the [magnitude] AC besides AB and BC that is continuous
with it. For what is taken from outside and is not a part of AC is not
continuous with it. It is the same argument, even if the motions of
5   the portions of the moving thing exceed the division of the motion
HI, so that the motions DE and EF, which are those of the portions
AB and BC of the [magnitude] AC,[57] are greater than the portions of
the motion HI, [with] which AC is hypothesized to be moving. For of
what [magnitude] is it the [corresponding] excess? Neither of the
whole [magnitude] (for the whole [motion] was [the motion] of this),
nor of any of its parts, for these [motions] are different from the
excess. If, accordingly, it is necessary that every motion be that of
10   something, but that neither the shortfall nor the excess of a [motion]
that falls short or exceeds will be [the motion] of anything, it is
obvious that [the motion] neither falls short nor exceeds, but rather
fits precisely, and the [motion] made of the portions is equal to the
whole [motion]. And if equal, [then] it is the same: for in fact all the
portions are the same as the whole thing. Having thus proved,
accordingly, that the division of the motion occurs according to the
15   division of the moving magnitude, since both the whole [motion] is
that of the whole [magnitude] and the parts are those of the parts,
he concludes the things that have been said. He said 'of everything
partitionable' either instead of 'of what moves' (for he proved that
every moving thing is partitionable) or he said 'of everything
partitionable' instead of 'of all the portions of a moving thing'. If the
motion is that of all the portions of a moving thing, it would also be
20   divided according to the portions of the moving thing.
    Eudemus [fr. 103 Wehrli] proved what was proposed [for proof] in
this way as well: 'For if there is a motion of the whole [moving
thing], and there is [a motion] also of each of its portions, and these
[latter motions] are different [from the whole] and each is a part of
the whole [motion], [then] the motions of the portions will be
portions [i.e. fractions] of the whole [motion], each in the proportion
that it itself too [i.e. the portion of the moving thing] stands to the
25   moving thing; so that if the portions of the moving thing equal the
whole [moving thing], the motions of the moving thing will also be
equal to the whole [motion].' Eudemus also says that there seems to
be a certain puzzle concerning what has been said: 'How ought one

[56] Inserting *sunekhes einai*, suggested by Diels in the apparatus criticus.
[57] Reading *tou AC kinêseis* with Aldine, as suggested by Cecilia Trifogli, instead of
tês AC kinêseôs with Diels and MSS.

to say that the portions move? For they do not exist actually, or else all partitionable things will be many and infinite, and nothing will be one.[58] If, then, one must [rather only] think the parts, what was stated [i.e. that the portions move] seems somehow possible in the case of transformation: for a leg grows white and [so does] each of the other [parts of the body]. The transformations of these seem numerically individual: for it is possible that one [part] grow white, and another not grow white. Let the transformations be the same in type [e.g. all to white]: the [transformation] of all [the portions] also becomes, accordingly, the same as that of the whole [moving thing]. For what is the difference in saying the [transformation] of each [portion] one by one or of all [the portions] simultaneously? But in the case of things that move locally [*pheromenon*], how shall we speak [of it]? For each of the portions has moved locally [with] an equal [motion], both the first that is taken and the last, and anyone whatever. But, then, the portions are infinite. Must the motion of the whole [moving thing], accordingly, be called one stade long or many stades long [i.e. the sum of the motions of the portions]? For in this way, indeed, the whole will have traversed infinitely many stades. Increase seems more in accord with transformation, so that the motions of these [i.e. increasing] things will be somehow partitioned according to the motions of the portions. But one must consider [further] concerning local [motion].' Eudemus, then, poses these puzzles in these very words. One must say in reply to them, I believe, what was said shortly before, [namely] that even if the parts of a continuous thing do not move locally in their own right, they nevertheless do move in the whole thing, at all events. And for this reason both the motion and the magnitude are said to be divisible, but not to have been divided. For they would have been divided if each of the parts of the whole [magnitude] moved in its own right.

30

974,1

5

10

235a10-13: 'There is another [division] according to time' to 'it is necessary that every motion be divided according to time.'

15

[Motion is also divisible according to the time taken. For every motion is in time, and the smaller the time the smaller the motion.]

Having said that every motion is divisible in two ways, in one way by time, and in another according to the moving magnitude, and having proved in three ways that it is divisible according to the

---

[58] Reading query after *kineisthai* and raised stop after *outhen* with Wehrli, rather than the reverse with Diels.

moving magnitude, he proves concisely and clearly that motion is
20   divisible also 'according to time'. For every 'motion is in time, and
every time is divisible'. Something moving evenly will move half [the
magnitude] in half of the time, a third in a third, and in general
more in more [time], and less in less. If there is more motion in more
time, and 'less in less [time]', and according to the same ratio, every
motion, consequently, is divided according to time. Alexander says:
25   'The division of motion that occurs according to time would be as
though according to length [i.e. in one dimension], while the
[division] according to the parts of the moving thing [would be] as if
according to breadth [i.e. in two dimensions]; for in fact time
proceeds according to a line that exhibits no breadth, while a moving
thing that has breadth moves as though on a surface, and not as
though on a line.'

975,1   235a13-24: 'Since every moving thing moves in something' to
'and again the lesser [motion] in the lesser [time].'

[Since whatever moves (or changes) does so in some respect,
and for some time, and motion belongs to the whole moving
thing, there are the same divisions of the time, motion, moving,
the moving thing, and the respect in which the motion (or
change) occurs, e.g. place or quality (although quality is
divided only incidentally because of the division of the body to
which it belongs). Thus, let A be the time in which a thing
moves, and B the motion; the thing moves with less motion in
less time, and vice versa.]

Having proved that all continuous things, [that is,] the moving
magnitude, and that in respect to which the motion [occurs]
(whether locally, qualitatively or quantitatively), as well as the
5   motion, the moving and the time, are divisible into forever divisibles
and that none is made of indivisibles, he proves here that there are
the same divisions of all these things and that all are similarly
co-divided with one another. The [expression], 'and there is a motion
of every [*pantos*] thing', is either short of [the words] 'moving thing',
so that it would be 'there is a motion of every moving thing', which
indeed is what he said earlier, [namely] that every motion is [the
motion] of something [234b29], [i.e.] of some moving thing, obviously
10   (for both that in respect to which the motion [occurs], and the time
in which it moves, are [the magnitude and time] of this [i.e. the
moving thing]); or [it means] 'there is a motion of all [*pantos*] the
thing' because all of a thing that is moving in its own right moves,
since it is [necessarily] a whole thing, and the [phrase] 'in

something' indicates that in respect to which the motion [occurs],
[namely] in respect to a place or a quantity or a quality. Having said,
accordingly, that the division of all these things is the same, since
quality seems not to be divided in its own right [by] a division into 15
parts, because it is not quantity, he added, therefore, that the things
in respect to which a motion occurs will not all be divided similarly.
For place, in respect to which local motion [occurs], and magnitude,
in respect to which increase and decrease [occur], being quantities,
will be divided in their own right, but quality, in respect to which
transformation [occurs], is no longer divided 'in its own right but
rather incidentally'. For by virtue of the fact that the body, in which
there is the quality, is divided, the quality itself will also be divided. 20
Having put down that the division 'of place is in its own right', he did
not further add 'and of magnitude', in respect to which motion
according to increase and decrease [occur], but, omitting this as
familiar, he added: 'but of quality incidentally.' Alexander says: '[We
may reply that] "of quantity in its own right" must have been
written instead of "of place in its own right", for thus he would have 25
spoken about both place and magnitude, for each is a quantity.'
Having added these things, accordingly, after this [Aristotle] proves
them: first, that motion is co-divided with time, and conversely, that
time is co-divided with motion. For if something moves [with] the
whole [motion] B in the time A, it is obvious that the thing moving 976,1
evenly will move [with] half of the [motion] B in half of the time A,
and the motion too will always be co-divided with the division of the
time according to the same [parts]. Conversely, if the motion is
divisible, [then] also in whatever way it is divisible (for this, above
all, is being proved here), it is necessary that the time too be
divisible in this way. For if something has moved [with] the whole
motion B in all the time A, it will move [with] half of it [i.e. the 5
motion] in half [the time] and always [with] the lesser [motion] in
the lesser time.

235a25-34: 'In the same way moving [*to kineisthai*] too will be
divided' to 'for if moving is taken with respect to each [motion],
the whole is also continuous.'

[Moving (*kineisthai*) is similarly divided, for there is less of it
with a lesser motion (*kinêsis*). If DC and CE are the motions
corresponding to the parts of the moving, C, then the whole
motion will correspond to the moving as a whole; nor can there
be a remainder (see above, on the motions of the parts of a
moving thing).]

10   Having proved that time and motion are divided similarly to one
another, he proves that moving too, which is an activity or affection
[*pathos*] in respect to motion, is divided similarly to motion.[59] For if
moving is the presence of motion, it is obvious that there is more
moving by virtue of the presence of more motion, and less [moving]
by virtue of the presence of less [motion], so that moving will be
15   co-divided with the motion, of which it is the presence. For if C is the
whole moving, and AD is the motion, the presence of which is the
moving, C,[60] with half the motion AB of the [motion] AD the moving
will be less: for it is half of C. And if we take half of the motion AB,
with this [motion] there will be half of the half-moving. And if we
forever divide the motion in this way, we shall divide the moving too
20   by the same division. In this way, then, he proved that if the whole
moving occurs by the presence of the whole motion, the parts of the
moving too will be divided according to the same parts of the motion.
Next he proves the converse through an illustration, that if the
portions of a moving have been co-divided with the presence of
portions of the motion [i.e. for each portion of the moving there is
present a corresponding portion of the motion], the whole moving
25   too will occur by the presence of the whole motion. For let the whole
motion DE have been divided at C, and let FG be the activity and
moving of the part DC of it [i.e. the motion DE], and let GH be the
moving of the other part, CE, of the motion, so that FH becomes the
whole moving in respect to the whole motion DE. If someone should
30   say that FH is not the moving of the whole motion DE, but either
that one of the [parts of the moving] FG [or] GH is, or some other
[moving] besides these, he will not speak rightly. For the motion DC
was that of the moving FG, and the [motion] CE that of the [moving]
977,1   GH; the whole [motion] DE was that of neither [of these]. But if [DE
is the motion] of some other [moving], there will be many movings in
respect to the same motion, which, indeed, is impossible. For of one
motion, numerically, there is one moving. He said, 'just as we proved
that a motion too is divisible into the motions of its parts', because,
having proved shortly before that a motion is co-divided with the
5   parts of the thing moving in accord with it, he said there too that if
the motion composed of the parts of the whole [motion] were not that
of the whole moving thing, but rather some other [motion] were, the
absurdity will follow that there are many motions of the same
[moving] thing. For the portions of the first motion that was taken

---

[59] A thing moves by the presence of motion (*kinêsis*); moving (*to kineisthai*) is the
activity of the thing that is affected by the presence of motion. Moving may, then, be
thought of as the state or condition of being in motion; 'movement' catches the sense,
provided that the verbal force of the term is borne in mind.
[60] Reading *hês parousia to* C *to kineisthai* with the MSS, instead of *hês parousiai*
(Diels) *tôi* (A, Aldine, Diels) C *to kineisthai*.

[i.e. that composed of the parts of the whole] will be the motions of it [i.e. the moving thing], and [so too] will be the [portions] of the second hypothesized [motion], if there should be another [motion]. Thus it would simultaneously be moving [with] the two [motions]. Having proved that, if someone should say that the whole moving does not occur by the presence of the whole motion, by the presence of the parts of which the parts of the moving were supposed to occur, [then] the consequent is absurd, he says in what way the whole moving will be the presence[61] of the whole motion. For if the moving was taken with respect to each portion of the motion, which is continuous, this [i.e. the whole moving] too would be proved to be continuous in a way similar to the motion, since it occurs according to it [i.e. the motion].

235a34-235b1: 'In the same way length too will be proved to be divisible' to 'it will be similar for all.'

[Distance is divisible in the same way, as is anything in respect to which there is a change (though in the case of quality, the division is incidental – see above); finite or infinite divisibility is again alike for all.]

Just as time, he says, and moving were proved to be divided by the same divisions, so too that in respect to which the motion occurs will be proved to be co-divided with the above mentioned, whether the motion [or process (*kinêsis*)] is in respect to place, ([i.e.] length), or in respect to quantity, such as increase and decrease, ([i.e.] magnitude), or in respect to quality, ([i.e.] quality): for whichever of these [latter] may be divided, all the others too [i.e. time, moving, etc.] will be co-divided by the same division. For if something should move [with] some whole motion in some whole time, whether in respect to place or quantity or quality, and if the time has been divided into a half or a third or any other portion [i.e. fraction], the motion too and that in respect to which the motion [occurs] will be divided into the same portions. Quality, however, if the motion should be in respect to quality, will not be divided in its own right, but rather incidentally (for [it is divided] not insofar as it is a quality, but rather insofar as what is being transformed or is changing in respect to quality is a quantity).[62] For it is by virtue of the fact that half of this [i.e. what is being transformed] is transformed in half the time that that in respect to which the

[61] Reading *parousia* with the MSS, instead of *parousiai* with Diels (in accord with the construction with *eimi* as opposed to that with *gignomai*).

[62] Closing the parenthesis after *estin* instead of after *poiotês*.

30   transformation occurs is also divided in half, for the [quality] in half
      of the whole is half of the quality in the whole. The moving thing too
      will be divided in a way similar to the abovementioned things, as
      [Aristotle] said in the first [parts of this book, cf. 234b10ff.]; for if it
      is [moving] as a whole in the whole time, half of it will be moving in
978,1 half [of the time]. But, while in the case of things moving locally by
      force [i.e. contrary to their natural motion], the moving thing is
      co-divided with the time – for, by force, a greater magnitude moves
      the same interval in more time and a lesser [magnitude] in less
      [time] – in the case of motion that occurs according to nature, how is
5   this sound? For, conversely, in things moving according to nature,
      the part moves in more time than the whole and the lesser [in more
      time] than the greater [e.g. larger things fall faster]. [We may reply
      that] in the case of these things too, that the magnitude is divided
      proportionally to the time and the time to the magnitude is
      preserved, but with the proportion taken inversely and not in the
      same way. For the division of a magnitude that is moving according
10   to nature will increase the time according to the proportion of the
      division. For by however much the thing moving according to nature
      is less, by so much the greater is the time in which it moves. The
      division of the time, again, will increase the magnitude according to
      the proportion of its own division; for by however much the time is
      less, by so much the greater is the magnitude moving according to
      nature [with] the same motion as a lesser [magnitude], so that the
15   proportion is inverse in the case of things [moving locally] according
      to nature and things moving locally contrary to nature. Having
      proved that all the things that have been mentioned are divided
      similarly to one another, he says that they are similar also in the
      matter of the portions that arise out of the division of them being
      finite or infinite. How the infinite is in them, and that it is according
      to [divisibility] to infinity, he will articulate in what [comes] next.

20   **235b1-5**: 'That [all things] are divided has followed above all' to
           '[that] the infinite [inheres] will be obvious in what follows.'

           [Divisibility and infinity belong above all to what is changing.]

           Since it is left to prove that what moves, which indeed has already
           been proved to be divisible, is also divided in a similar way to the
           others, he proves for good measure that all the other divided things
25   are divided in consequence of what moves. For neither the time nor
           the motion nor the moving nor the place or magnitude or quality of a
           whole moving thing and of a portion of it are equal. Division is,
           reasonably, a property of the other things as a result of this [i.e.

what moves], because the rest are [properties] either of this or of the [properties] of this. For length is [a property] of a body, as are motion and magnitude and quality, while time is [a property] of motion. Quality, however, has [the property of] being divisible more 30 self-evidently as a result of the magnitude that is moving [i.e. changing (Aristotle has *metaballontos*, 235b3)]; for a quality is neither continuous nor divisible in its own right [cf. above, 977,25-30]. Having added that all things have the infinite in a similar way, beginning with the magnitude that is moving [or changing], he says that it has been proved that all things have divisibility in a similar way. For this is what he proved immediately before. How the infinite is in them and how it is similarly in all, will 35 be proved later. Alexander, again, inquires how motion in respect to 979,1 transformation has its division according to the division of the magnitude that is moving, if indeed some things change all at once, too, as [Aristotle] himself said in the first [book] of this treatise, referring to Melissus [cf. 186a15]. Alexander also remarks here that even if not all but [only] some part of it changes all at once, neither 5 the motion nor the time would any longer be co-divided with the division of this portion, and he decides [the matter] by saying: '[We may reply that] it is not co-divided with all the portions of it, but rather with the [portions] of it as a moving thing, insofar as it moves.'

235b6-19: 'Since everything that is changing changes from something to something' to 'for it is similar in the case of one [change] and the others.'

[As soon as a thing has changed, it is already in the position or state into which it has changed, since to have changed from something is to have left it behind; this applies to change between contradictories, and hence to all changes.]

It is proposed to prove that it is necessary that what has changed, as 10 soon as [literally, 'when first' (*hote prôton*)] it has changed into that into which it was changing, be in that into which it has changed. For it is not simply necessary that what has changed already be in that into which it was changing. For if something changing from black to white should be in grey, this will have changed from black, but will not yet have changed, however, into white, into which it was 15 changing, on which account it is not yet in white, into which it was changing. The [word] 'first' [*prôton*] is added to 'it has changed' because in this way the statement will be true. For it is necessary that someone coming to Athens be in Athens as soon as he has come;

however, it is no longer necessary that someone who came [literally, has come] to Athens last year be in Athens now: but he was then, as
20   soon as he came [cf. Themistius 193,11-15]. Such, accordingly, is the problem. Having assumed it in advance as self-evident that everything that is changing 'changes from something to something', he proves what was proposed by a transfer of terms and by the fact that change in respect to coming to be and perishing self-evidently possesses this [property]. The transfer, then, was from 'changing' to 'leaving behind'. For [Aristotle] says that 'either leaving behind is the same thing as changing', or invariably 'leaving behind follows
25   upon changing'. If these [terms] are such with respect to one another that either they are the same or one follows upon the other, it is obvious that having changed and having left behind will also be such with respect to one another that either they are the same or
980,1   having left behind follows upon having changed. With this having been assumed in advance, it is evident, in the case of change in respect to coming to be and perishing, which is [a change] from not-being to being, that to have come to be is to have changed from not-being, and to have changed is to have left not-being behind. What has left not-being behind is invariably in being, for there is
5   nothing between the contradictory pair of being and not-being; and a change into being was happening to it. Having proved what was proposed in the case of coming to be, he added that if it is thus in [the case of] this change, it is also true generally in all [changes] that everything that has changed, as soon as it has changed, is in that into which it was changing. This is true by the hypothesis that
10   states that what has been said is similar both in the case of any one kind of change soever and in the case of the other [changes].

        235b19-32: 'Further, it is evident even to those who take the changes] one by one' to 'it is obvious that, as soon as it has changed, it is in that.'

        [What has left behind that from which it has changed must be somewhere: if not in the position or state to which it has changed (call it B), then in C; but since change is continuous, it would then still be changing from C to B, and then it would not be the case that it had changed.]

Having ruled that what has changed is in the case of the other
15   changes too such as it is in the case of coming to be (for what has changed, then, is in that into which it was changing), he attempts here to prove that this same thing is so also in the case of each change, and he uses a general proof that is able to fit the case of

every change. At the same time, by means of this demonstration he also resolves an objection brought against the preceding demonstration. For someone might say that what has changed, even if it has left behind that from which it is changing, might not yet be in   20 that into which it is changing, but rather in something in between: for in most changes there is an in between. And one ought not to prescribe, on the basis of changes according to contradictory pairs, which do not have anything in between, for other changes as well, in which there is an in between. He resolves this objection too, accordingly, by this argument. For if, he says, 'it is necessary for   25 what has changed to be somewhere' – then, lest someone not[63] understand [the word] 'somewhere', as it is stated, in reference to some condition, but rather in reference to a place, which indeed is what 'somewhere' seems to mean, he added 'or in something', which indeed seems more general – if, accordingly, it is necessary that what has changed be 'in something', whether a place or magnitude or state or either part of a contradictory pair, and it has left behind   981,1 that from which it was changing, either it will be in that into which it was changing, or, if [it is] in any other thing that is before that into which it was changing, it would still be changing into that into which it was said to have changed and would not have changed [into   5 that]. For without a [process of] change from that in which it is hypothesized to be it will not be in that into which it was supposed to be changing, for there will invariably be some interval from this to that; for these are not partless things adjacent to one another, in such a way that it is possible to have changed from this to that apart from [a process of] changing. For if the change from A to B, into which it was changing, is continuous, it is necessary, if it is not yet in B, that it be changing toward it; for it was supposed to be   10 changing into this. But if the change toward B is not continuous, [then] neither was it changing into this originally, but rather into that in which first it stopped its change; so that if what has changed is not yet in B, it would still be changing into it (into which, indeed, it was [originally] changing) which indeed is absurd. It is impossible, consequently, for what has changed to be in anything   15 other than in that into which it was changing. Alexander says: 'It is possible to prove, by [the statement] "the change is continuous", that what has changed from A into B will not be in any other thing, either, that is outside the [change] AB and separate from it. For if the change is continuous, and a continuous [change] is on a continuous thing and does not leave a gap, [then] something that is separate from it would not be continuous with the [change] AB.' Having resolved, accordingly, the [above]mentioned objection as   20

---

[63] Inserting *ouk* after *pou*.

well, [Aristotle] concludes that it has now become evident, too, that what has been said is true also in the case of change according to contradictory pairs: for 'what has come to be, when it has come to be', will then be, and when it has perished, then 'it will not be', and 'in general, concerning every change', what has been said has been truly said, even if it is rather more manifest in [change] according to contradictory pairs. Since it is necessary that what is changing

25   according to any one of these changes be in either [one or the other] part of the contradictory pair, but it is impossible that it still be in that from which it has changed and [which] it has left behind, it is necessary, consequently, that it be in the other.

982,1   **235b32-36a7: 'In which it has first changed'[64] to 'have the one perished, the other come into being, in an atomic [now].'[65]**

[What has changed has done so in an indivisible now. For if it has changed in the divisible time AC, divide AC at B; if it has changed in AB and BC, then it has not changed in AC first or primarily; nor has it changed in AC if it is changing in AB and BC, or has changed in one but is changing in the other.]

Having proved that it is necessary that what has changed, as soon as it has changed, be in that into which it was changing, he proves here that this 'in which first what has changed has changed' is not

5   time, but rather some atomic [i.e. indivisible] limit of time, which indeed we call a 'now', but which Plato called a 'sudden' [*exaiphnês* (cf. *Parmenides* 156D)]. Having explained what 'first' means, he passes thus to the demonstration of what is proposed. He says that in general a particular thing or sort or amount, and similarly for all [categories], is said to be 'first' [or 'primary'] which 'not by virtue of some other thing before it being such' is itself also said to be such.

10   Opposed to 'first' [or 'primary'] is 'by virtue of something else'. What is said to be [so] by virtue of something else is either what is said to be such as a whole because some part of it is such, just as we say that seeing is not a property of a man first, because [it is a property] of the eyes first, and we are said to be in Athens by virtue of the place in which we are being a portion of Athens; or else, what is itself also said to be such because its kind [or genus (*genos*)] and that

---

[64] Aristotle's text reads, 'in which first it has changed' (*prôtôi* in place of *prôton*), which is the way Simplicius himself cites the phrase in 982,4, 20, etc.
[65] Aristotle here uses the word 'atomic' (*atomon*) in the etymological sense of 'uncuttable' or 'indivisible'; the reference is to a sizeless instant rather than to an extended 'atom' of time.

under which it has been classified are such,[66] as an equilateral 15
triangle is said to have its three angles equal to two right [angles]
not first, but rather because every triangle [does so]. Since,
accordingly, something can be said to have changed in a certain
time, because it has changed in some one of the [parts] of this [time]
(for [it has changed] in a limit of it), for this reason he added the
[word] 'first'. He proves that 'that in which first what has changed 20
has changed' is atomic by reduction to the impossible. For, if it is
possible, let that in which first it has changed be divisible, let it also
be AC, and let it have been divided at B. It is necessary, accordingly,
that what is hypothesized to have changed in AC first have changed
in each of the [parts] AB and BC, or else to be changing in each [of
the parts], or to be changing in one [of the parts] and to have 25
changed in the other. But if it has changed in each of [the parts] AB
and BC, it has no longer changed in AC first, if, at all events, it has
changed also in the first portion of it, [that is,] AB. How too could it
be said to have changed in both first, since when it has changed in
the first part, it is not yet in the second, while when [it has changed]
in the second, it is no longer in the first? But if it is changing in both 30
portions of AC, [that is] AB and BC, it will also be changing – not 983,1
have changed – in AC as a whole. But it was hypothesized to have
changed in AC first; for what is changing in all the parts of
something would also be changing in it as a whole. If it is still
changing in one part of AC, but in the other it has changed, there 5
results an absurdity, [namely] that there is 'something earlier than
the first'; for it was hypothesized to have changed in AC first. If,
accordingly, it has changed in either part whatsoever of it, it would
have changed in this first, and then, if indeed [at all], in the whole,
so that something will be taken [as] earlier than the first. This is
absurd both in its own right, [namely] that there is something
'earlier than the first' (for that earlier than which there is something 10
is no longer first), and, in fact, because 'in [something] first' was said
to be [precisely] that, [namely] not [being] in a part of it but rather
in the whole. If, accordingly, what has changed was in a part of AC,
AC was no longer the first in which it has changed. Having proved,
accordingly, that upon its being hypothesized that that in which
first something has changed is divisible, there follows something
impossible with respect to each segment of the [logical] division, he
reasonably concluded the argument by saying: 'thus, that in which it 15
has changed would not be divisible' – in [which] first, obviously, even
if he omitted it as clear.
  Since he proved in the argument before this that what has
changed, as soon as it has changed, is in that in which it has

---

[66] Omitting Diel's insertion of *ho*.

changed, and that this appears self- evident above all in change in
20　respect to coming to be and perishing, he reasonably added here too
that 'it is evident that both what has perished and what has come
into being have the one perished, the other come into being, in an
atomic [now]'. For if 'what has perished and what has come into
being' is [equivalent to] 'the one perished, the other come into being,
are in that in which first [each] has changed', and [if] 'the one
25　perished, the other come into being are in that in which first' is
[equivalent to] 'they have the one perished, the other come into
being in an atomic [now]', [then] 'what has perished and what has
come into being have the one perished, the other come into being in
an atomic [now]'. 'By this,' says Alexander, 'he has indicated to us
also the solution of the argument posed by the sophists, [namely] at
what time did Dion die: for it was either when he was living or when
he was dead; but not in [the time in] which he was living (for at that
[time] he was living) nor in [the time in] which he was dead (for he
was dead in all that [time]). If, accordingly, Dion died neither in the
30　time in which he was living nor in that in which he was dead, Dion
did not die. But [the word] "died" means the first [now of] change
from living, while [the words] "when he was dead" [mean] the time
in which he was dead. And if what has changed has not changed in
984,1　time, one need not fear the [conclusion], "Dion never died"; for he
never [did die] in the sense of [dying] in time, since it is in an atomic
[now] that he has changed from living.'

236a7-27: ' "In which first it has changed" is said in two ways'
to 'for the divisions are infinite.'

[Something is said to have changed first in two senses: when
the change is completed, and when it begins; but there is no
first time at which a thing begins to change, or is changing. For
let AD be the first time at which something begins to change;
AD is divisible, since if what changes was at rest in the earlier
time CA (and if CA is indivisible), the thing is at rest also at A;
but it has changed at D, so AD must be divisible. The thing
must, then, have changed in any and every part of AD; and,
since the divisions are infinite, there is no first.]

5　Having proved both that what has changed, as soon as it has
changed, is in that into which it has changed, and that that in which
first it has changed is atomic (for it is not in time, but in a now), he
adds, in consequence of these things, that 'the first' in which
something has changed is spoken of in two ways: that to which what
is changing was changing, in which it both has stopped from

changing and is [in a state of having] changed in it; and also that 'in   10
which first [what is changing] began' its change and [the process of]
changing is said [to be the first]. For this too is called that in which
first it has changed, because, I imagine, it switched from not
changing. With 'that in which first it has changed' thus spoken of
and understood, accordingly, in two ways, he says that that which is
called first in respect to the end of a change both is and is called [so]
in the strict sense. For the change has been completed, and there is   15
something in which first it was completed, as has been proved,
[namely], a now. For the end of a change is not [itself] a change. But
what, at the beginning of a change, which beginning is itself also a
change, is called the first [bit] of the time in which the beginning of
the change [occurred] – this, he says, is not in existence at all, but is   20
spoken of and understood emptily, because it is not possible to take
some first [bit] of the time in which what is changing began its
change. For it is not possible to take a beginning either of the change
or of the time in which something first was changing, because the
beginning of a change is itself also a change, and what is in it is not
outside the [process of] changing, as [is the case] in the limit, when it
had changed. The proposition would be that 'there is no beginning of
a beginning', since whatever is hypothesized as the beginning of the   25
time will be discovered to have another beginning.

He proves this also through an illustration, as follows: for, he
says, let some first [bit] of time be taken, in which what is changing
began its change, and let this be AD. AD would not, then, be
indivisible. For time will [then] be composed of indivisibles; for the
beginning of a change is a change, and every change is in time, so   30
that if AD is indivisible by hypothesis, time [too] will be, and time
will be composed of indivisibles, which indeed is what [Aristotle]
meant by 'it will result that nows are bordering [each other]'. But he   985,1
said that this results upon its being hypothesized that AD, in which
it began its motion, is indivisible, because it will result that motion
occurs in a now, since AD, being indivisible, is a now, and in this
[now] it began to move, and the beginning of a motion is a motion. If   5
something should be moving in a now, since every moving thing
moves in time, nows would be parts of time and time would be
divided into nows and composed of nows. For that in which motion
can occur is a portion of time. That AD is not partless he proves
through another argument too, as follows: what begins to move in
AD, begins invariably to move as it changes from rest. Accordingly,   10
let it be at rest in CA and beginning to move in AD. If, accordingly, it
is at rest in all of CA, it is obvious that it is at rest for a bit also in A,
which is [a part] of CA. If, accordingly, AD is partless, AD will be the
same thing as A. Consequently, that which was beginning its
change in AD would also be at rest in it. But what is beginning its

15   change in it is moving in it. Consequently, it will be at rest and
moving in the same thing, which indeed is impossible. He took being
at rest in A as equal to not moving. For it was supposed that in a
partless [now] it is possible for something neither to move nor to be
at rest. What follows then is not, in truth, being at rest and moving
20   simultaneously in the same now, but rather not moving and moving
simultaneously, which indeed is [even] more impossible on account
of the axiom of contradiction.

Having proved that time at the beginning of a change is not
indivisible, but rather divisible, he proves next what was proposed,
[namely] that it is impossible in this [case] to take anything [i.e. any
bit of time] in which first what is changing changes: for if AD is not
partless, it is necessary that it be divisible and that it not change in
25   it first. For if it changes in AD first, it will either be changing in any
portion whatsoever of it, or in neither [of two parts], [or in both],[67] or
in one or the other. But if in both [of two parts] or in one or the other,
it is not changing first [i.e. primarily] in the whole. And if it is
changing in neither, it would again not be changing in any way
whatever in the whole. If, accordingly, it changes in the whole, it
changes in any portion whatsoever of it. But there will be many
portions, since the whole is continuous, and every continuous thing
30   is divisible to infinity, so that it is impossible to take that [portion] in
which first it changes. For it is similar in every [portion] that is
taken, and there is not a first in a division to infinity. The argument
is about first changing (for the beginning of a motion was [said to be]
a motion), but [Aristotle] himself uses 'having changed' instead of
986,1   'changing', since what changes has changed when it reaches the end
of changing.

It seems puzzling, in these [matters], how a limit both of motion
and of time is said to exist, at which what is changing is said to have
changed, but there does not exist a beginning. For in fact
5   Theophrastus [fr. 25 Wimmer] in the first [book] of his *On Motion*
says: 'Amazing things are seen to exist in respect to the nature itself
of motion,[68] for example, if there is no beginning of it, but there is a
limit.' How did [Aristotle] take the limit [to be] indivisible, but the
beginning divisible to infinity? For it is possible, using the same
arguments, to take both the limit of a continuous thing to be
10   divisible to infinity, and the beginning to be indivisible. It seems,
accordingly, both that limit is of two kinds and that beginning is of
two kinds, whether of motion or of time or of any continuous thing,
on the one hand as the first or last part of the continuous thing, and,
on the other hand, as beginning and limit, which are no longer parts

---

[67] Inserting *ê en amphoterois*.
[68] Omitting *ta*, inserted by Diels before *tês kinêseôs*.

or similar to the whole. For in the case of such things it was ruled
that a beginning and that of which it is the beginning are not the
same thing, nor a limit and that of which it is the limit, just as a
point is both the beginning and the limit of a line but is not itself a   15
line, and similarly a now [is a beginning and limit] of time and a
move of motion: for so they call the limit of a motion. And at first he
was saying that what has changed was in such a limit, [i.e.] one
which is after the whole change. But here he takes the beginning
both of [a stretch of] time and of a motion as a part. It is obvious   20
from the fact that he calls this the beginning 'in which first it began
to change', that is, to move. Every moving thing moves in time, and
not in a limit of time. Such a beginning, being part of a continuous
thing and [itself] continuous, is divisible into forever divisibles. Such
a thing does not have a beginning, because whatever [part of it],
indeed, you may take, that too is divisible, since the indivisible
beginning, at all events, which is opposed to the limit in which a   25
thing is said to have changed, is such that nothing changes in it, as
neither [does anything change] in the limit, but rather it exists
before the change and does not have another beginning before
it[self]. [Aristotle] himself, taking on the one hand a beginning of
motion and a first motion in respect to changing as a part of motion
and as being in time, and, on the other hand, a limit as that in which   30
first what is changing has changed, says that the latter exists and
can be taken ('for it is possible for a change to be completed'), while
the former does not exist nor is it possible to take it, because of every
beginning in the former [case] there is another beginning, just as
also in the case of a limit [taken] as a part it is possible to take
another limit of every limit. And it is obvious that if the whole is   987,1
taken as a finite continuous thing, it is possible to take both partless
limits, and endings [*teleutaia*] [taken] as parts, but when it is taken
as divisible to infinity, it is not possible to overtake either partless or
partitionable limits, for before every partitionable thing that is
taken there is another, and after every one another. Even if some   5
first or last part should be taken, since this is partitionable, it is
necessary that it have changed in any [part] whatever of it, so that
there will again be something earlier than it, and, since what is
taken is always partitioned, only that which is a beginning or limit
in the sense of atomic will be first.

236a27-36: 'Neither is there some first thing of that which has
changed' to 'it is evident from what has been said.'   10

[There is no first bit DF of a thing DE that has changed, for
everything that changes is divisible; if HI is the time in which

DF has changed, in less time some bit of DF will already have changed.]

Having proved that it is impossible to take any [bit] of time in which first what is changing began its change, he proves here that neither is it possible to take some first thing, in the sense of a part, of the changing thing itself. For here too he takes such a first thing, and not the partless thing that is opposed to the partless limit, which 15 indeed, in the case of time, he called that in which it has changed. And he proves this too through an illustration, using a reduction to the impossible. For since every changing thing is divisible, let FE be the changing thing as a whole and let it have been divided into FD and DE. And of DE [a mistake for FE][69] let FD be the first that has changed. Let HI be the time in which FD has changed; for every 20 change is in time. And observe that here he was making HI the time in which FD has changed, not because he took 'has changed' as a limit, but rather because he put it down instead of 'was changing'. If, accordingly, DF [order of letters as in Aristotle] has changed in all the time HI, [then] in half of the time HI, since every time too is divisible, something less than DF will have changed, and also, 25 obviously, before DF as a whole, and again of this portion some other thing first, in accord with the division of the time that is taken. Thus of a first there will always be another first, and it will be impossible to take any [bit] of what has changed which has changed first. Upon this, then, he concludes that it is not possible to take some first thing either of a changing thing or of the time in which the changing thing is changing.

988,1   **236b1-18**: 'The thing itself which is changing or in respect to which it is changing' to 'it is possible to be indivisible in its own right.'

[That in respect to which a change occurs, whether place, quantity, or quality, is divisible, save that quality is divisible only incidentally; thus in none of these is there a first bit. For example, let the magnitude AB have moved (locally) from B to C first: BC cannot be indivisible (else partless things will touch); thus the magnitude has changed or moved into something prior to C.]

Having proved that there is neither any first bit of the time in which

---

[69] In Aristotle's text, DE (rather that FE) designates the changing thing as a whole; Simplicius has accidentally copied Aristotle's words without adjusting the notation.

what has changed began its change, nor is there a first bit of the
changing thing itself which has changed, since every change is of 5
something, [namely] of what is changing, and in something (for it is
in time), and in respect to something (for it is either in respect to
place or quantity or quality or a contradictory pair), he has shifted to
this third thing, [namely] that in respect to which the change occurs,
inquiring in this case as well whether there is some first thing of
that in respect to which a change [occurs]. Having said about this, at
the beginning, 'the thing itself which is changing', he substituted for
it the clearer [phrase], 'in respect to which it is changing'; for he 10
added the latter as explanatory of the former. And he says that what
is being inquired into in the case of everything in respect to which a
change [occurs] is no longer similar to [the object of inquiry] in the
case of the changing thing itself and the time [in which it changes].
For the latter are divisible by their own nature, and on account of
this it is not possible to take any first [bit] of them in which the
beginning of the change [occurred]. But everything in respect to 15
which a change [occurs] is no longer divisible in its own right. For
quality, in respect to which change by transformation occurs, is not
divisible by its own nature, because it is not a quantity. It too is
incidentally divisible, however, by virtue of the fact that what
underlies it [*hupokeimenon*] and is transformed is divisible.[70] For
the quality too is co-divided with this, since it too is itself quantified
incidentally; for being divided is [a property] of quantity. He said 20
that all things are divisible incidentally, not as though place and
magnitude are also divided incidentally (for these are divisible in
their own right), but either he was saying that all qualities in
respect to which transformation [occurs] are divisible incidentally,
or that, if what is divided incidentally is also counted along with the
divisible things, [then] all things in respect to which change [occurs]
become divisible; and it is true in the case of all things that it is not 25
possible to take a first [bit] on account of division to infinity. Then, of
the things in respect to which change occurs, taking the things that
are divisible in their own right, and first [of all] place, he proves in
the case of this that it is not possible to take a first [bit], illustrating
AB as the moving magnitude, and the place on which it moves
having itself B as its beginning, and C as its end, because a moving 30
thing begins its motion when it has touched the beginning of the
place with its own beginning or end. Therefore he took B in common
when he took BC as a portion of the place, which indeed must be
either indivisible or divisible. And if B and C are indivisible and

---

[70] In a transformation from black to white, the underlying thing does not become
white all at once, but rather some bits become white sooner than others, and because
the underlying thing is infinitely divisible, there will be no first bit that is
transformed, and thus no first bit of white; cf. 989,16-30.

989,1 bordering one another, the place BC, being continuous and having
magnitude, will be composed of indivisibles, and partless and
indivisible things will be bordering one another, which has been
proved to be impossible. If the limits of the place, B and C, have been
divided from each other, there will be something before C, for
example D, into which first what is moving in respect to place from
5 B toward C will have changed, and again another thing earlier than
D, for example E, and of that again another first, and so [on] forever,
so that it is impossible to take a first [bit] of the place into which the
moving thing has changed, because of the fact that the division
never fails. The argument is similar also in the case of change of
quantity by increase and diminution. He also added the cause of the
10 similarity: for it is because the increasing magnitude is both
continuous and divisible into forever divisibles. For this much [of a]
magnitude increasing in this much time will increase less in less
[time], and again less in a [time] less than this. Because he produced
the earlier proof for the case of place, he indicated there one thing by
taking AB as the moving magnitude, and another [by taking] BC as
15 that on which the moving thing is moving from B to C. The
increasing and diminishing magnitude is in the same [form].

Having proved that it is impossible to take a first thing either of
place or of magnitude on account of their being continuous and
divisible into forever divisibles, he concludes that of those things in
respect to which motion [occurs], only quality can be indivisible in
20 its own right. But it is obvious that this too is divisible –
incidentally. Even if it is indivisible by its own nature, there will not
be a first thing in it either (for a first [bit] is in a thing that is divided
and has parts); but rather there is both no first thing of the
underlying thing in respect to which [the change occurs], with which
[the quality] is co-divided, nor would there be any first thing of the
co-divided thing itself. A transformation is co-divided unless the
25 change should occur all at once, which [Aristotle] himself, arguing
against Melissus in the first book [of the *Physics* (186a15-16)], ruled
[able] to happen, when he said: 'as though the change were not all at
once.'[71] Eudemus, however, put it more cautiously concerning this
[matter] in the fourth [book] of his Physics [fr. 104 Wehrli], writing
as follows: 'How ought one to take things that are transformed as
changing? For if a body is transformed as a quantity, and does not
30 all grow warm or dry simultaneously, there will not[72] be a first thing

[71] At 966,19, this passage is quoted with *ginomenês*, 'occurring' (in accord with the
MSS of Aristotle) in place of *ousês*, 'being'.
[72] Reading *oude en toutois* with MSS CM, rather than omitting *oude* with Diels and
Wehrli. Simplicius is citing Eudemus' more cautious formulation of Aristotle's own
view; the caution lies in explicitly recognizing the case of the all-at-once change of a
part, at least, which requires another argument. (For *sômata* as the subject of
paskhousi, cf. Bonitz' index to Aristotle s.v. *numerus* ad init.)

in these [cases], while if bodies undergo changes in another way [e.g. if their parts change all at once], [that is] another argument.'

236b19-32: 'Since every changing thing changes in time' to 'it is   990,1
necessary, consequently, that it have moved in any [part]
whatever of XR.'

[The exact amount of time a thing takes to reach the
completion of its change can be called the 'first' time in which it
changes, and it is changing in any part whatever of this time;
for let a thing be moving in the time XR, and divide XR at K;
only if it is moving in both XK and KR will it be moving in XR
'first'.]

Having proved that it is impossible to take a first [bit] in the time in
which something changes, [then,] since something is said to change
in a first time if it is [changing] not in some part of the said time but
in it immediately [*prosekhôs*], [Aristotle] proposes to prove that this   5
[sense of] first too, being divisible, does not admit of taking a first in
the strict sense. For changing in time is said in two ways, either as
'in a first [or primary time]' or as 'by virtue of something else', and
something changing in one of the parts of this [i.e. time] is [a case of]
'by virtue of something else'; for a contest [occurs] in this year,
because it is on this day of it [i.e. the year], but [it is] on this day, or   10
in this hour, as a first [time]. How, accordingly, is this time said to
be first, if indeed it is not possible to take a first time? He proves that
this [time] too is not first in the strict sense, since this too is
partitionable, since it is time, and it is necessary that what is
changing change in any part whatsoever of it, so that it is not
possible to say in what [part] first what is moving began its motion.
For of the thing that is taken as first it is possible for some other first   15
[bit] to be taken, and of that another, and so [on] forever. He proves
also from the definition of a such a first that being some part of a
whole it too is divisible. For something was said to be or move in a
first time which was not, or was not moving, in any one of the [parts]
of the said time; for 'first' is opposed to 'by virtue of some part' and to
'by virtue of some other thing'. This time too, accordingly, has been   20
taken as partitionable, since it is first on this account, [namely] that
it is not in any of its parts [which parts thus exist by definition].

He produces a third demonstration again by an illustration and by
reduction to the impossible, that neither in this case is it true that
there is properly a first time in which something moves. He takes,
then, XR as the time in which first the moving thing moves. And
since 'every time is divisible', he divides it at K. The thing, then, that   25

is moving in the [time] XR first is 'either moving or not moving' in the part XK of it, and similarly in KR. And if it is moving in neither, [then] neither will it move in the whole [time] XR, but rather will be at rest. For it is impossible for what is moving in none of the portions
991,1   of something to move in the whole thing. But if it is moving in one of the portions of it, [either] XK or KR, and is not moving in the other, it would not be moving in XR first, for it occurs as moving in some one of the [parts] of this, which indeed was not [what was defined as] moving in the time XR first. For [moving] in a first [time] was [defined as] not [moving] in some one of the [parts] of this, but
5   rather in all of it. The time XR is not, consequently, first, since every time is divisible. But if 'in the first time' was this, [namely] not in some one of the [parts] of this, but in all of it, it is possible that what is moving in all of the time XR is moving in it first. For even if the time XR is partitionable, what is moving not in some one of the [parts] of it but rather in all of it might be said to move in it first. But
10   it is not possible to take [it as] a beginning of time and first in this way. For whatever part of time you take as first, it has something earlier than itself on account of the division of time to infinity. But it is possible to take a certain time as first if it is taken as a whole together. For in a whole [time] a thing altogether changes, and not in a portion of it, so that the change is coextended with the motion [i.e. passage] of it [i.e. the time]. For a contest is completed in this
15   particular month, but not [in it as] first, but [in a] first [time] on this particular day and in hours of this particular day, in all of which [hours] alike it is completed. And this time is first, not because there is not another time before it, but because there is not another [time] before it in which the contest is completed, and because [it is completed] not in any part of it whatsoever, so that the division is
20   carried on to infinity, but rather the whole contest is completed in the whole [time], and the change of it is coextended with the motion [i.e. passage] of the time. Aristotle proved, however, in the illustrated demonstrations that it is not possible to take a first of time in the sense of the beginning of a time. Therefore, he also used the division of time to infinity, nowhere taking it [i.e. time] as a
25   whole thing. But next, taking it also as whole, he calls it a first time, when he says: 'for if KL has moved in the first time XR' [236b34- 35].
        Since the problem is basic, I think it is not absurd to cite also what has been written on this by Eudemus, who understood the thought of Aristotle better than all his other commentators. In the fourth
30   [book] of his *Physics* [fr. 105 Wehrli], Eudemus writes as follows: 'How, accordingly, will a first in which it was changing be possible? For it is possible in that in which [the change] was completed; for there is a limit in this. The end or the first are said in several ways: in one way, in that in which as in an atom; this is possible in the case

of what has changed, since it is possible to have changed in an atom
[of time, i.e. a now], but it is not possible in the case of being at rest
or changing, since it neither moves nor is at rest in an atom [of   992,1
time]. It is also possible in that in which first it moves, as has been
said, but not, in fact, in some [part of the time], but rather in all of it
in a similar way. For we say that the sun was in eclipse in the month
of the Olympics, because [it was so] on the first day of the month;
and on the first day of the month, because in some portion of this
[day]; but [we] no longer [say that it is] in this particular portion,
nor in another, but rather that it is in any [part] whatsoever of this   5
[time].[73] But this is partitionable, so that [the eclipse occurs] earlier
in an earlier [part of the particular portion]; for a part in the part is
earlier. But the whole [eclipse] is in the whole time, so that [the sun]
was in eclipse in this first as in a [stretch of] time. There is,
accordingly, no first time, since all [time], in which first it was
moving or undergoing anything else, is partitionable, since such
things are in time and in some part [of time] in such a way as to   10
occur in any [part] of it whatsoever. This is what is called "in a first
[time]".'

236b32-237a2: 'This having been proved, it is evident' to 'so
that the moving thing will have moved.'

[Everything that is moving has moved earlier. First proof: if
the magnitude KL has moved in the time XR first, something
moving at the same speed will have moved less in less time; so
too, therefore, the original moving thing.]

It having been proved that every time, and also that called first, is
divisible to infinity, he says that it follows upon this that every
moving thing necessarily has moved earlier – the moving thing is
said to have moved not in the sense of having stopped its motion, but   15
rather in the sense that the traversing thing has already traversed
some amount. For since a moving thing, whenever it is moving, has
undergone a change [*metêllakhen*] from something to something (for
if it were in the same thing, it could not be moving), and what has
undergone a change has moved, it is obvious that it is necessary that
the moving thing have moved. He himself proves it as follows: he
takes KL as moving in the first time XR. It moves, accordingly, in   20
every portion of it: for he takes [the term] 'it has moved', as I said,
not in the sense that what is said to have moved has already stopped

---

[73] Reading *oute en allôi* with MSS and Wehrli, instead of *hoti en allôi* with the
Aldine and Diels.

its motion, but rather in the sense that it has already traversed some amount in its moving. Since, accordingly, there is some time XR in which first the moving thing KL changes by changing in all of

25   it, if some other thing is taken, moving at a like speed as KL, and beginning its moving simultaneously with KL, in half the time of XR, that is in XK, it will traverse half [the magnitude]. In the same time [i.e. XK], KL, which is equal in speed to it and started simultaneously with it, will also traverse the same [magnitude]. For what moves in all of it has moved in any portion of the time whatsoever. Consequently, what moves [over] the whole [line] has

30   moved [over] half of it earlier. Since KL was hypothesized to have

993,1   moved in the first time XR, for this reason he took some other thing, of a like speed as KL, and moving in half the time of XR half of what KL moved in the whole time XR, so that through the former he might prove also that KL itself had also moved the half before having moved the whole. For if he took KL itself, in its own right, as having moved in half of the time XR, [that is] XK, he would seem to

5   be subject to a certain objection, because KL was hypothesized to have moved in the first time XR.

237a3-11: 'Further, if in all the time XR' to 'every changing thing will have changed an infinite [number of times].'

[Second proof: any time XR in which a thing has moved is bounded by nows, between which there is always a divisible stretch of time; hence, whatever is changing has changed an infinite number of times.]

This is a second argument proving that it is necessary that every

10   moving thing have moved earlier. For if we say that everything that has moved has moved in any time whatsoever, by taking the last now that bounds the past time in which it was moving (for before every now there is time because nows do not border on each other, and every time, even the least, is divided by a now), it is obvious that even in that now which divides the past time in half it will have

15   moved, so that what is moving toward the last now of the time, before moving [with] this motion, had moved, when it traversed the half. For there was also some last now of the half, and not only of the half, but also of any greater and lesser part whatsoever, because every time, even the least, is divisible. And it is divided at the now,

20   and at each now it has moved just as at each time it is moving. If this is so, and every time is divided to infinity by nows, and between all nows there is time, so that nóws do not border on each other, the

moving thing not only has moved before moving, but it has also moved an infinite number of times. For as many times as it is possible to take a now on the time in which the moving thing moves,   25 just so many times is it possible to predicate having moved of it. It is possible to take a now just so many times as it is possible to divide the time. The time is divisible to infinity, as he will prove. Consequently, the moving thing will have moved an infinite number of times. For if every time is divisible, and what is between the divisions is also time because nows do not border on each other, this [time] too is divisible; so that if every time is divisible, and after   30 every division that has been taken time is left over, the division will never fail by virtue of the fact that, within the [successive] cuttings, what is able to be divided never fails. But this is what time was [shown to be like]; consequently, the division is to infinity. To whatever extent there is division, to that extent too there are nows. For the division is at these. To whatever extent there are nows, to   994,1 that extent too having moved is predicated of the moving thing. Consequently, the moving thing will have changed an infinite number of times. It is obvious that [it will have changed] an infinite number of times potentially and not actually, just as the division too has the infinite potentially by virtue of the divided thing being [able to be] divided to infinity, but not by virtue of its having been at some   5 time divided to infinity in actuality.

237a11-17: 'Further, if a continuously changing thing' to 'every changing thing will have changed an infinite [number of times].'

[Third proof: it is not possible for a thing to be changing in a now; hence, it must *have* changed at each now. But the nows are infinite in number.]

This is a third argument proving that every changing thing has changed earlier to infinity. For it is necessary for a continuously changing thing, which has neither perished nor stopped its motion   10 to any extent whatever, either to change or to have changed in the time in which it is said to be moving. But, in fact, it has been proved that in a now it is not possible to change. Consequently, it has changed, so that at each of the nows that are taken in the time in which the moving thing moves continuously, it will have moved and have changed. But the nows are infinite in the sense of [division] to   15 infinity; consequently, the changing thing will have changed an infinite number [of times].

237a17-28: 'Not only a changing thing' to 'so that it would be changing earlier.'

[Whatever has changed was changing earlier. First proof: if it has changed from A to B it is no longer in A; thus it is at B in a different now. But there must be time between the nows, during which it was changing, and likewise before any now at which the time is divided.]

20 Having proved that it is necessary that every changing thing has changed earlier, he proves also that its converse is true, [namely] that it is necessary that everything that has changed be changing earlier. He proves this by demonstrating in advance that everything that has changed has changed from something to something in time. This he proves by reduction to the impossible. For if it has changed from A to B not in time, but in the same now, it will be simultaneously in A and in B, which indeed is impossible, for it is

25 not possible for what has changed, when it has changed, to be in that from which it has changed, as he proved earlier, when he said that changing and leaving behind are either the same thing or leaving behind follows upon change [cf. 235b9-11]. The statement would have been clearer, if to the [phrase], 'for it would be simultaneously in A and in B', he had added, 'which indeed is impossible', and then had mentioned the [statement] of the impossibility, saying: 'for it

30 has been proved earlier that it is not possible for what has changed,
995,1 when it has changed, to be in the same thing.' If, accordingly, it has not changed in the same now from A to B, but rather in different ones, and between every two nows there is time, since nows are not bordering,[74] [then] it is obvious that it endured the change in time; for this is what 'to have changed in time' means here, and not that

5 [sense of 'to have changed'] which is a limit of changing. If, accordingly, it has changed in time, and every time is divisible, it is obvious that it has also changed half in half of the time, and [has changed] half before, at all events, [having changed] the whole, and that in half of half the time it was changing again half of half the magnitude, and so [on] forever on account of the division of the time to infinity. It is not possible for what has changed by portions to

10 have changed as a whole all at once, but rather it will proceed through changing to having changed. For what has changed would not have changed through changing only if the change were occurring atemporally and the magnitude were composed of atomic things. Since every change occurs in time, it is necessary for

[74] The term 'bordering' (*ekhomenon*) embraces 'continuous' and 'touching'; cf. 925,27-926,1.

everything that has changed from something to something to have   15
changed through changing earlier.

237a28-b9: 'Further, in the case of magnitude it is more
evident' to 'exactly as in the case of lines that are increased and
reduced.'

[Second proof: if something has changed from C to D, CD
cannot be indivisible (else the partless things C and D will
border on each other); CD is thus a magnitude and infinitely
divisible, and hence what has changed was changing earlier.
Therefore, changing always precedes having changed, and vice
versa.]

Having proved from the time in which a change [occurs] that before
having changed there is changing, he says that the same thing is
more self-evident from the magnitude in respect to which a change
[occurs], obviously in the case of things changing in respect to   20
magnitude. These are things changing in respect to place and in
respect to increase and diminution. For things changing in respect
to quality are not subject to this demonstration from magnitude,
just as neither are things that come into being and perish. For the
change of these things is not in respect to quantity, so that it is not
in respect to a continuous and forever divisible thing either, from
which [properties] the demonstration arises. It is obvious that what   25
is proposed is more evident in the case of magnitude than in the case
of time, since in the demonstration from time he used in addition
magnitude as well, when he said: 'in half [the time] it will have
changed another [change], and again another in half of that [time]'
[237a26-27]. But in the [demonstration] from magnitude he does not
need time. He proves it by taking two magnitudes, either of place, in   30
the case of change in respect to place, or [of that] in respect to which
increase and diminution [occur], and positing C and D [respectively]
as that from which it changes, and that to which it first changes. In
these cases, it is necessary that CD as a whole be either indivisible
or divisible. But it is impossible that it be indivisible, for a partless   996,1
thing will be bordering a partless thing, [namely] that from which
and that to which. But if it is divisible, there is a magnitude in
between, of which that from which is the beginning, and that to
which the end. A magnitude is divisible into forever divisibles, 'so
that it is changing earlier' into the parts and into the parts of the   5
parts. Having said this, he draws the conclusion in common [to all
the arguments] that 'it is necessary that what has changed be
changing earlier'. Since the immediately preceding demonstration

from a magnitude or a continuous thing did not address things changing in respect to quality, which he called contraries, nor in respect to coming to be and perishing, which he calls 'in a contradictory pair',[75] he says that in these things too there is the
10   same demonstration, taken from the time in which it has changed. The demonstration before this one was of such a kind, making common use of the continuous and forever divisible [stretch] of time in the case of every [kind of] change. He next, then, reasonably stated both general conclusions together, [namely] that it is necessary both that 'what has changed be changing' earlier and that
15   'what is changing have changed', and each always before the other. For it is not possible to take a first and a beginning of a change on account of the fact that continuous things are divisible to infinity and are themselves not composed of partless [entities]; for division in the case of continuous things is infinite, and this is obvious, he says, 'exactly as in the case of lines that are increased and reduced'. For if we take any line and divide it in two, and if we keep one half
20   indivisible, while dividing the other half of it again in two, and if we add one [half] of it [i.e. the divided half] to the half that was [kept] indivisible from the beginning, and if we forever increase the [half kept indivisible] from the beginning by the addition of halves of the divided [half], the reduction and the increase will be shown to be to infinity.

237b9-22: 'It is evident, accordingly, that it is necessary also that what has come into being come to be earlier' to 'so that, in
25   whatever it is, it would not be [in it] as in a first.'

[A continuous thing that is coming to be or perishing must have in part come to be or perished, and vice versa.]

Having said previously that the same demonstration that proves that there is changing before having changed pertains also to things that are not continuous – [that is], to contraries and things according to contradictory pairs – when the demonstration is taken from time, here he proves that also in the case of things that change
30   according to contradictory pairs – that is, things [that change] from not-being to being or from being to not-being, that is, of the things that come to be and perish as many as are divisible and continuous – not only will the demonstration from time fit, but also that from

---

[75] i.e. the preceding demonstration applied only to local and quantitative change, the second of which he thinks of as being a change between contrary states. It did not apply to a substance's switching between existence and non-existence, which he thinks of as a change between contradictories.

magnitude, so that 'it is necessary that a magnitude that has come   997,1
into being be coming to be earlier, and that one that is coming to be
have come into being'. And the argument is true not only in the case
of things that are coming to be, but also in the case of all things
changing in any way whatsoever. Not, indeed, that what is coming
to be has itself always also come into being earlier, but rather some
one of its parts [has]. For if a house is still coming to be, some [part]   5
of it, [i.e.] the foundation, has already come into being when the
walls and the roof are coming into being. And perhaps it is true to
say in the case of the foundation itself that something of it as it is
coming to be has always come previously into being, on account of its
being partitionable to infinity.

Having said, 'but [some] other [part of it has come into being]',
why did [Aristotle] add, 'sometimes'? For in the case of something
still coming to be some part of it has always come previously into
being. [We may reply that] 'sometimes' is added not in those cases in   10
which having come into being precedes coming to be, but rather in
those cases in which coming to be [precedes] having come into being.
For, in fact, sometimes this very thing was said to be coming to be
which is now said to have come into being, for example a house or a
foundation, and sometimes what was coming to be earlier was not
this thing which has now come into being, but rather was some
other thing and part of it [i.e. of what has come into being]. For if a
house has come into being, it was coming to be earlier; but it was
coming to be both when the entire house was being completed, and
not [only this], indeed, but also when its foundation was coming to   15
be, what was coming to be was some [part] of the house.

He says that 'it is similar also in the case of something that has
perished' to the way it is in the case of something that is coming to
be. For, in fact, having perished precedes perishing, and perishing
[precedes] having perished. Also in the case of these it is not
necessary that what has perished be that which was perishing
earlier, nor that what was perishing be that which has perished. But
rather a portion too of the whole can be what precedes, for as it is in   20
the case of coming to be, so must it be also in the case of perishing.
And as the cause of the fact that these things [mutually] precede
each other in the case of coming to be and perishing, he added that a
certain infinity inheres in things that come to be and perish. For
coming to be and perishing properly and in their own right are
properties of bodies, and bodies are divisible to infinity. The reason
for this is that they are continuous: for every continuous thing is   25
divisible to infinity, because continuous things are not composed of
partless things. The time, too, in which there is the coming to be and
perishing, being continuous, is itself also divisible to infinity, so that
each will always precede the other both in the case of coming to be

and in the case of perishing on the basis both of the demonstration
from time and on the basis of that from magnitude.

30   Alexander says that the [phrase] in the beginning of the passage,
[namely] 'as many as are divisible and continuous', is added because
there are some things, in the case of which we predicate that they
came to be, although they do not have their being [*einai*] by virtue of
coming to be [*genesis*]. We say that contacts [*haphê*], at least, have
998,1 come into being, [but] not, indeed, that they are coming to be.[76] And
it is obvious that in such cases having come into being does not
precede coming to be nor [does] coming to be [precede] having come
into being,[77] since coming to be is not in them at all. In general, that
changing always precedes having changed, and having changed
5   [precedes] changing, is more familiar in the case of change in respect
to place because the interval of place is continuous and divided
similarly to time, while, in the case of other changes it is safe to
draw the distinction that, if all of the change is in time, [then] it is
true that changing precedes having changed, and having changed
10   changing, in respect to time. But if certain changes also occur all at
once, then perhaps one ought no longer to admit to this – all at once,
[that is,] not only as, in general, the portions [of a thing may change
all at once], as in the case of freezing, but rather [all of the change
occurring] not as in time. For in these cases changing and having
changed do not precede each other, because there is no changing at
all in the case of them. Theophrastus [fr. 54 Wimmer] seems to be
puzzled whether every change is in time, suspicious, perhaps, of
15   changes from darkness to light, when a lamp is brought into a
chamber and all of the room is filled with light all at once, without
[an interval of] time [cf. Themistius 197,4-8]. But contact, if it does
not have its being by virtue of coming to be, is not said to have come
to be either, but rather to be without coming to be. Therefore, in the
first [book] of *On the Heavens* [280b6-9], Aristotle says that one of
the meanings of 'ungenerated' [*agenêtos*] is what [occurs] according
to contact.

20   **237b23-238a18: 'Since every moving thing moves in time' to 'it
would move [over the interval] AB in a finite time.'**

[A thing cannot move over a finite line in an infinite time,
whether or not the speed is uniform. Let AB be a finite interval,
AE a portion of the interval that measures it off, and CD the

[76] Contact is not a process; if two things are coming together, contact arises just
when the change has occurred (i.e. is completed). Similarly, if a line has been divided,
the two segments are in contact, without having come into contact.
[77] Following the word order of F and the Aldine, instead of Diels'.

infinite time. AE is traversed in a finite time, since it is less than AB and must therefore be traversible in less time; the sum of the finite times in which all the parts equal to AE are traversed will be finite.]

It is proposed for him to prove that it is neither possible to move [over] a finite line or [with a finite] motion in an infinite time nor [over] an infinite [line or with an infinite motion] in a finite [time]. First, [he proves] the first thing mentioned: he distinguishes in advance in what sense, if a finite line or motion is taken, it is true that it is impossible to move [over] a finite [line or motion] in an   25
infinite time, so that he says[78] that it traverses the whole [line], which is finite, once in all the infinite [time], and not the same [line] many times, as we see occurring in the case of the revolving body [of the heavens], or some [part] of the same [line] again and again; for in this way it is possible [to move over a finite line in an infinite time]. Thus, it is impossible to move [over] the same [line] once in an   999,1
infinite time. He has already proved before that it will neither traverse things finite in multitude in an infinite time, nor things infinite [in multitude] in a finite [time], when he objected to the argument of Zeno [cf. 233a21-b15]. He proves this same thing here first in the case of magnitude, then also in the case of motion itself,   5
using a more general demonstration. He assumes in advance certain of the things that have already been agreed upon, [namely] that every moving thing moves in time, and that the same thing moves [over] a greater magnitude in more time and, obviously, [with] more of a motion as well. Since the moving thing moves either at a uniform speed or at a non-uniform speed, in the case of a thing moving at uniform speed, first, he proves what was proposed, in   10
which case the demonstration is also easier. For if a certain portion is taken of a whole line or a whole motion 'which will measure off the whole' (which is finite), there will be some finite [portion] of the time, as well, in which it was moving [over] the portion that was taken of the line or the motion. For if it was moving [over] the whole [line or with the whole motion] in an infinite time, it is obvious that [it was moving over] the part in a finite [time], by virtue of the axiom   15
assumed in advance, that says that the same thing moves [over] a greater [magnitude] in more time. Accordingly, in finite times, equal in number to 'however many portions there are' of the line and the motion, the moving thing 'has moved [over] the whole' [line and with the whole motion]. If, accordingly, the parts of the line and the motion have been delimited both in magnitude and in number (for

---

[78] Omitting *gar* with CM; Diels in the apparatus criticus suspects a corruption in the text.

20   the [phrase] 'each [is delimited] in how much and in how many times' means this for [Aristotle]), the portions of the time too would have been similarly delimited both in magnitude and in number: that of which the portions have been delimited both in magnitude and number would itself as a whole also have been delimited [i.e. finite]. Consequently, the time in which the moving thing will traverse a finite line and motion is not infinite, but finite.

25   Even if the moving thing does not move at a uniform speed, he says, it makes no difference in regard to proving that the moving thing will traverse a finite magnitude and a finite motion not in an infinite, but in a finite, time, if one [assumption] only is kept, [namely] that in more time a moving thing gains in addition another interval, whether it moves at uniform speed, or whether it abates

30   and intensifies [in speed] by turns, or whether it always remains in one or the other of these, either in abatement so as always only

1000,1 [further] to abate, or in intensification so as always [further] to intensify. If some time has been added in which a thing that is moving in any way whatever has moved some amount,[79] invariably some interval too, either of magnitude or motion, will be added to the interval which it has already moved in the time that was taken before. And the time too is again divided, even if into [parts] unequal to [those of] the line and the motion, yet into [parts] finite, at all

5   events, in number and magnitude. He proves it by assuming first what has been proved already, [namely] that everything that moves has moved earlier [over] another portion of this [interval] in a portion of all of the time. For of the portions of the magnitude and the motion [over or with] which it moves, the moving thing has moved [over] one [portion] earlier than another, producing an order of the motion according to the first and the later portions of the time.

10   For it has not moved [over] all the interval simultaneously. Therefore, since the motion occurs in this way, if [in] a particular amount of time a thing moving either at a uniform speed or at a non-uniform speed has moved some particular amount [of a line or motion], [then] when some time has been added it will move more [of a line or motion] in more [time]. For [it will] certainly not [move over] the same [line or with the same motion] both in more [time] and in less, if indeed that axiom is true that says that a moving

15   thing moves [over] a greater interval in more time and not the same [interval] in more [time] and in less, since both the motion and the magnitude have been added relative to an addition of time. Therefore he took it that the moving thing moves [over] the first portions of the magnitude or [with the first portion of the] motion in

---

[79] Reading *to...kinoumenon* with F (as suggested by Cecilia Trifogli), rather than *tou...kinoumenou* with Diels and the other MSS; mention of part of the moving object is irrelevant here.

accord with the first portions of the time. If, accordingly, the interval and motion, for example AB, have portions, of which the moving thing has moved some first and some later, if we subtract some   20
portion AE which will measure the whole finite [line or motion] AB, and [if] we subtract from the infinite time CD [the portion] in which the moving thing moved the former portion at non-uniform speed, a portion too of all of the [time] will obviously be finite. And if we again subtract another [portion] of the interval and the motion, and take the finite time of this, even if it is not equal to the first time that was   25
subtracted, it is nevertheless finite, because the moving thing was moving [over] the whole finite interval in an infinite time and it is impossible to move [over] both the whole and the part in the same [time].[80] If, then, by forever subtracting from the finite interval or magnitude or motion [portions] equal to the first that was subtracted we exhaust it, and [if] for each [such portion] we subtract   30
from the infinite time CD a finite [time] in which the thing moving at non-uniform speed is moving, there will be as many portions of the number [*arithmon*, 'number', is a mistake for *khronon*, 'time'] as there are of the magnitude, with those of magnitude finite in respect to number. For every finite thing is divided into [portions] equal to one another that are finite in number. Consequently, the times too,   1001,1
in which the moving thing will traverse all of the interval, will be finite with respect to number. The time, being composed of [portions] finite in number and interval, will itself also be finite. Consequently, in a finite, and not an infinite, time, the moving thing will move both a [over] finite line and [with a finite] motion. For even   5
if the intervals of the time are not equal because of the non- uniform speed of the motion, nevertheless, being finite, at all events, in number, and each finite in magnitude, they make the whole finite, whether they are equal or unequal. The syllogism would be in the second figure[81] as follows: The time in which a moving thing moves   10
[over] a finite line or [with a finite] motion is exactly completed and measured off by portions finite in number and magnitude; an infinite time is not exactly completed by such portions; consequently the time in which the moving thing will traverse the finite [line or motion] is not infinite. Since it was not possible for there to be taken any portion of an infinite time which will measure it off, and for each portion of the magnitude [over] which the moving thing moves, or of   15
the motion, some finite portion of the time can be taken, and the portions of the time are co-completed exactly with the portions of the magnitude or motion, whether they are equal to each other or

---

[80] Reading a comma after *to kinoumenon* instead of Diels' full stop, and a full stop after *kineisthai* instead of Diels' comma.

[81] See n.6.

unequal, and time made of [portions] that are finite both in magnitude and in number is finite and not infinite – on account of this he handled the demonstration in this way.

20   **238a18**: 'Likewise also in the case of resting.'

If 'resting' [*eremêsis*] signifies for him [i.e. Aristotle] rest [*êremia*] as opposed to motion, since what is at rest too is at rest in time, he would say that just as it is impossible for a finite motion to occur in an infinite time, so too for rest. But if resting signifies also that a motion that is toward [i.e. coming to] rest is being brought to rest, which indeed in the next [passages] he calls 'halting' in the sense of
25   coming to a halt [238b24], then he reasonably rules the same things to be so also in the case of resting as [he does] in the case of motion. For having cut motion in two, that relative to the thing from which a moving thing is moving he calls 'moving', but that relative to the thing to which [it is moving he calls] 'halting' and 'being brought to rest'.

1002,1   **238a18-19**: 'Thus it is not possible for something that is one and the same either to be coming to be or to be perishing forever.'

He inferred a certain corollary from what has been said, [namely] that one and the same thing does not forever either come to be or
5   perish, that is a finite thing in an infinite time. For this is what 'forever' means. For if every finite motion and change occur in a finite time, and coming to be and perishing are [a kind of] changing, every finite coming-to-be and perishing occur in a finite time; and the coming-to-be and perishing of a finite thing are finite. For just as
10   it is impossible to move [over] a finite line in an infinite time unless it is again and again, so too it is impossible for a finite thing to come to be and perish in an infinite time.

        238a20-31: 'The same argument [proves] also that not in a finite time' to 'for it will be the same argument.'

        [It is not possible to move or be brought to rest over an infinite line in a finite time, whether at a uniform speed or not. For in a part of the time that measures off the whole time, a finite part of the magnitude will be traversed.]

Having proved that it is impossible to traverse any finite magnitude, or for there to be a finite motion, in an infinite time, here he proves 15 the converse of that, [namely] that it is impossible to traverse either an infinity, or for there to be an infinite motion, in a finite time. But neither [is it possible] to be brought to rest for an infinite [period of] resting [i.e. coming to rest] in a finite time. Even if 'being brought to rest' means 'being at rest' as the opposite of moving, he would say that neither is it possible for something to be at rest for an infinite 20 rest in a finite time. If the [phrase] 'moving neither evenly nor unevenly' has been added to 'moving' and 'being brought to rest' in common, [then] 'being brought to rest' would mean 'heading toward rest', but if it has been added relative only to 'moving', [then] 'being at rest' would rather be meant by 'being brought to rest'. He uses the same demonstration now too, dividing the finite here also and co-dividing the infinite with it, except that earlier, having divided 25 the magnitude (for it was the thing hypothesized as finite) he co-divided the time which was hypothesized as infinite, while here having hypothesized the time as finite and subtracting some portion from it which measures the whole, since in each of these times the 1003,1 moving thing moves some finite [amount] (equal [amounts] when it moves at a uniform speed, unequal when not at a uniform speed, but always finite; for it will not, then, move an infinite [line or motion] in a portion of the whole time; for it was hypothesized to move an infinite [line or motion] in the whole time), the magnitude hypothesized as infinite will also be co-exhausted, accordingly, with 5 the finite time, whether the segments should be equal or unequal. Given that the division is finite, in a time that is finite the division in the magnitude too will similarly be co-limited 'both in how much and in how many times', that is, both in magnitude and in number. What is made of [parts] finite both in magnitude and in number is not 10 infinite, so that it will not traverse an infinity in a finite time. He adds that taking the magnitude as infinite 'on one side' or 'on both' makes no difference in regard to the proof (it is obvious that even earlier, [taking] the time [on one side or on both made no difference, cf. 950,27-951,3]). For even if it is taken on both, the same proof proceeds. For it is necessary that, in a finite portion of a finite time, some finite portion of the magnitude hypothesized as infinite on 15 both sides be taken, and when the finite part has been taken as many times as in the case of the time, which is finite, whether the parts are equal or unequal, he will prove that the magnitude hypothesized as infinite is finite. Earlier too [233b11-14], having taken the time as finite on one side, he proved that a moving thing will move a portion of a finite magnitude in a finite time, even if it 20 should be possible to move [over] all [of the magnitude] in an infinite [time (cf. 950,21-2)].

238a32-b16: 'These things having been proved, it is evident' to 'further, even if the time is taken, there will be the same demonstration.'

[Since a finite magnitude cannot traverse an infinite magnitude in a finite time, neither can an infinite magnitude traverse a finite, for the two occur together. If A is an infinite moving thing, a part CD of it and then other parts successively will be alongside the finite thing B, so that each will traverse the other. Neither, consequently, can an infinite magnitude traverse an infinite magnitude in a finite time. The demonstration using time is the same.]

Having proved earlier, and again immediately before, that a moving
25   thing will move neither [over] a finite magnitude nor [with] a finite motion in an infinite time, and similarly the converse, that neither will it move [over] an infinite magnitude or [with an infinite] motion in a finite time, he draws as a corollary the consequences of these things, [namely] what results in the case where the moving thing itself is no longer taken in an indefinite way, as earlier, but rather is
30   distinguished, hypothesized now as infinite, now as finite. For, he says, with the abovementioned things having been proved, it is evident from the same demonstration in the case of the moving thing that if it should be hypothesized as finite, it is impossible that it traverse some infinite magnitude in a finite time, and if it should
1004,1   be hypothesized as infinite, if it should move an infinite [amount] wholly, it will neither traverse a finite nor still more an infinite [amount] in a finite time. That something which is finite will not traverse some infinite magnitude in a finite time he proves by the
5   same demonstration as previously, too: for if the finite time is divided into finite portions, the moving thing, in each [portion] traversing some finite part of the magnitude, exactly completes the whole, so as to traverse some finite and not infinite [magnitude] in a finite time. But if the moving thing should be hypothesized as infinite, and the interval through which it moves as finite, the
10   infinite thing will not move through a finite [interval] in a finite time. For making either the moving thing infinite and the interval finite, or vice versa, makes no difference, he says. For if an infinite [moving thing] A moves through a finite [interval] B, there will be some part of the infinite [moving thing] in B, for example CD, 'and
15   again another and so [on] forever', until all the parts of the infinite moving thing pass by B, so that it will result that simultaneously the moving thing, being infinite, has traversed through B and B, being finite, has traversed an infinity. For it is not possible to say that an infinite thing traverses a finite thing in any other way than

by the finite thing traversing through the whole infinite thing. For
an infinite magnitude, if indeed it should be infinite by being
everywhere, does not itself have anywhere to pass to, but may be   20
said to move through some finite thing only in this way, [namely] if
the finite thing should traverse entirely through all of its portions;
and if this, indeed, should happen, the infinite will be measured
back by the finite, for in no other way is it possible to traverse it. If,
accordingly, this is both absurd and the finite is again seen to be
traversing through the infinite here too, which was proved to be
impossible, it is obvious that an infinite thing could not traverse   25
through a finite thing in a finite time, and well he said that it makes
no difference 'which one soever the moving thing is [i.e. the finite or
the infinite thing]'.

Having proved that neither will a finite magnitude traverse an
infinite interval in a finite time nor an infinite magnitude a finite
interval, he proves here that neither will an infinite magnitude
traverse an infinite interval in a finite time. For if an infinite   30
[magnitude] as a whole were going to traverse entirely through an
infinite [interval], it is necessary that some finite part of it traverse
it earlier. And it is not possible to say that the infinite does not have
a finite part; for the finite inheres in the infinite. If, accordingly, it
has been proved that it is impossible for finite thing to traverse [an   1005,1
infinite interval] in an infinite time, it is obvious that neither will an
infinite thing traverse it. We shall prove the same thing, he says,
beginning from time as well. It has been proved earlier that it is not
possible to traverse an infinite thing in a finite time. For if we   5
subtract from the time some portion which measures off the whole
[time], which is finite, in this [time] it will move [over] some portion
of the infinite thing, since it was hypothesized to be moving [over]
the whole infinite thing in the whole finite time. If, accordingly, for
each of the portions of the finite time we take some portion of the
infinite thing, since the portions of the time, which is finite, are
finite, the [portions] of the infinite thing too will be exactly co-   10
completed, and it will no longer be traversing an infinite thing.

238b17-22: 'Since neither will a finite thing traverse an infinite
[magnitude]' to 'the other too [must] be infinite.'[82]

[Neither can there be an infinite motion in a finite time.]

It has already been proved earlier with the magnitude on which the

---

[82] The phrase 'for every locomotion is in place' (238b22) is omitted here; but cf.
1005,32.

motion [occurs] that a moving thing will neither move [with] an infinite motion in a finite time nor a [with] finite [motion] in an
15   infinite [time]; for the demonstration was [of both] in common. For however large the interval through which the motion [occurs], just so large is the motion as well. Therefore I too took motion along with magnitude in the demonstration, [Aristotle] himself having said in the beginning of the theorem that 'it is impossible to move [sc. either 'over' or 'with'] a finite one [feminine adjective] in an infinite time, when not always moving [over or with] the same one
20   [feminine pronoun] and any of the [parts] of it [237b24-26].' He put down the feminine term perhaps because he was speaking about the line [feminine noun] through which the motion [occurs], or perhaps because he was considering the motion [feminine noun]. But if he ruled the same things [to be so] concerning rest too, which is the opposite of motion, [then] it is obvious that what was demonstrated earlier also concerned motion, for in fact his argument extended to this [latter]. Here, however, on the basis of what was proved concerning the magnitude through which the motion is accom-
25   plished, he infers that neither is the motion infinite in a finite time. The magnitude through which the motion [occurs] is local, and on this [magnitude] the motion in respect to place has its interval. For since, he says, neither an infinite nor a finite magnitude moving [with] a motion in respect to place will traverse an infinite interval in a finite time, 'it is evident that neither will the motion be infinite
30   in a finite time', unless it should occur again and again on the same [interval]. For every infinite motion in respect to locomotion [*phora*; cf. 243a8] occurs on an infinite local interval, for every locomotion is
1006,1 a motion in respect to place. And the infinite motion is one [motion] on an infinite place, unless it should be the same [motion] again and again. If, accordingly, it has been proved that it is not possible to move [over] an infinite interval in a finite time, it is obvious that neither is it possible [to move with] an infinite motion. For the interval too, through which the motion [occurs], would be infinite. But insofar as concerns the proof that there does not exist an infinite
5   motion, what function did [the proposition] that the moving thing too is not infinite serve? [We may reply] that if there is an infinite magnitude and it moves at all, even if for any amount of time whatsoever, the motion will be infinite inasmuch as it is co-extended with the magnitude of which it is [the motion]. For it has been proved that a motion is co-extensive with the magnitude, too, of which it is the motion, just as it is with the interval through which
10   the moving thing will traverse. Accordingly, he nicely co-eliminated the infinity of a motion in a finite time not only by means of the infinity of the interval through which the moving thing moves, but also by means of the infinity of the moving magnitude. For if the

motion is coextended alongside both of these, [then] if the thing moving in a finite time has neither of these as infinite, neither would it have an infinite motion.

238b23-39a10: 'Since everything either is moving or is at rest' to 'there will not be that in which first it halts.'   15

[Whatever is halting (coming to a halt) is moving rather than at rest; hence it halts, faster or slower, in time. Consequently, there is no first time AB in which it is halting; for AB is divisible, and the thing is halting in any part of it.]

What is proposed for him is to prove – even if he seems to be speaking paradoxically with respect to what is apparent – that what halts is the same as what moves, and that however many things are properties of what moves, just so many are properties also of what halts. Having taken what is at rest, [i.e.] that which is opposed to what moves and remains in the same [place], as one thing, and what   20
is brought to rest, [i.e.] that which moves toward being at rest, as something whitening [moves] toward being white, as another, he says that what has halted is the same thing as what is at rest, while what is halting [is the same thing] as what is being brought to rest, which indeed is what is coming to a halt. And first he proves that what is halting moves, assuming in advance a certain necessary disjunction that says that 'everything either is moving or is at rest', [everything] that is so constituted, obviously: for what is motionless   25
by nature neither moves nor is at rest, but what indeed is so constituted as to move is also [so constituted as] to be at rest, since being at rest is the privation of moving. What is so constituted as to have [a property] is deprived [of it] when it is so constituted as [to have it] but does not. For a recently born puppy, indeed, would not be said to be at rest with respect to not seeing, because it is not then so constituted as to see. Nor would a man who does not have eyes in   30
his chest be said to have been deprived of having [eyes] in his chest,[83] for he would be said to have been deprived where he was so constituted as to have [something] but did not have [it]. Neither would fish be said to have been deprived of motion in air, because they were not so constituted as to move there. Nor again have fish   1007,1
been deprived of motion by walking, for they are so constituted as to swim, but not to walk. It is necessary, accordingly, for everything that is so constituted as to move, if it is not moving when it is so constituted and where it is so constituted and in whatever way it is

---

[83] Omitting *mē* before *ekhein* in line 30, with MSS CM.

so constituted, since it is in [a state of] privation of moving, to be itself at rest. It is necessary, accordingly, for what is halting, that is
5 coming to a halt, since it is so constituted as to move, either to be moving or at rest. Therefore, it is not at rest, for what is already at rest is no longer being brought to rest, while what is halting is being brought to rest. For just as being brought to rest, so too halting is a route to rest, for both are the same thing. Therefore he substituted 'halting' for 'being brought to rest'. If, accordingly, what is halting moves, and what is moving moves in time, it is evident from what
10 has been proved that what is halting halts in time, since what is halting, being so constituted as to move and be at rest, is not at rest but rather is moving, and what is moving moves in time.

He proves this also another way: for if 'it is possible to halt faster and more slowly' (for what is coming to a halt will come away either faster or more slowly), and everything that is faster or slower is in time, halting consequently is in time.

Having proved that what is halting also moves and halts in time,
15 he proves next that the impossibility of taking a first [time] in which the halting thing halts is also a property of a halting thing, just as it is of a moving thing, because before everything that is posited as first there is something earlier, as was proved also in the case of motion by means of the cutting to infinity of the time and of the interval [over] which it moves and of the motion itself and of that in respect to which the moving thing moves, whenever it moves in
20 respect to quantity. That it is not possible to take a first in which the halting thing halts, but rather whatever time one takes as first, another will be discovered before it, he proves by proving earlier that in the time in which first it halts, 'it is necessary that it halt in any [part] whatsoever of this'. For when the time in which first the
25 halting thing is said to halt is divided, if it halts in neither of the parts, neither [will it halt] in the whole, so that the halting thing would not be halting; but if it halts in one or the other, and first, obviously, in the first, it would not be halting in the whole as first, for it halts in the whole in the sense of halting by virtue of either of the parts,[84] just as was said earlier in the case of a moving thing [cf. 236b19-23]. The moving thing and the halting thing are similar
30 because a halting thing too is a moving thing. Just as it is necessary for a moving thing, accordingly, in whatever first [time] it is hypothesized to move, to move in some one of the [parts] of this, so [too it is necessary] for a halting thing. Having proved this in advance, accordingly, he proves on the basis of this that just as it is

[84] The MSS of Simplicius and of Aristotle here are divided between *kath'hekateron* (Diels), which would mean 'in respect to (or by virtue of) either of the parts', and *kath'heteron* (Ross), which would mean 'by virtue of (or in respect to) something else', cf. Simplicius 990,8-10.

impossible to take a first, in which something moves, of moving, so
too neither [is it possible] of halting. He proves this too through a
reduction to impossibility. For, he says, let 'that to which AB' [is
applied as label][85] be that in which first it is said to halt. What he
earlier divided straightway, [Aristotle] here first proves to be
divisible, on the basis of its not being partless. He proves that it is
not partless on the basis of there not being motion in a partless
thing, so that neither is there halting, since halting is some [kind of]
motion. That there is no motion in a partless thing he recalled
through what had already been proved. For if it is necessary that a
moving thing have moved earlier, [then] there is no motion in a
partless thing. For it is not possible that something move and have
moved in the same thing, and 'a halting thing has been proved to be
a moving thing'. Since, then, AB, in which first the halting thing is
said to halt, is divisible, it halts in any of its portions whatsoever; for
this has been proved in advance, [namely] that in whatever first
something is said to halt, it is necessary that it halt in any of the
portions of that thing whatsoever. If, accordingly, 'it is time in which
first' it is said to halt, and not some atomic [i.e. indivisible (atomon)]
thing, and every time is divisible to infinity, there will not,
consequently, be any first [bit] of it, nor will there, consequently, be
any beginning of halting, in which the halting thing halts first. For
as is the time in which the halting [occurs], so too is the halting in it.
Accordingly, from the fact that it halts in any of the portions
whatsoever of that which is divided to infinity [i.e. time], he proved
that there is not any first thing in which it halts. For neither is there
any first thing of that which is divided to infinity, but rather, if
halting is moving, and in the case of moving it has been proved that
it is impossible to take anything in which first a moving thing moves
[cf. 1006,21-2], it is obvious that neither is it possible to take any
first thing of halting.

What is the function, accordingly, of the demonstration in this
case as well, to one who argues briefly [i.e. to Aristotle]? It is that
wishing first to demonstrate that this halting is some [kind of]
motion, even if the statement is contrary to custom, he reasonably
was led first to prove that however many things are properties of
moving, just so many are [properties] also of halting, inasmuch as
halting is some [kind of] moving. Then, if moving and halting were
indistinguishably the same, the argument concerning halting would
be otiose; but since this halting is not simply moving, but rather
moving toward a halt, since it is the same thing, as has been said, as
being brought to rest, he reasonably produced a demonstration in

[85] Simplicius here adopts or quotes Aristotle's formula *to eph'hôi AB*; normally he
omits the periphrasis and writes *to AB*.

30   the case of such a moving as well, so that he would have an
      argument that is true in the case of every motion, both that relative
      to the thing from which a motion [occurs] and that relative to the
      thing to which [it occurs]. And that Aristotle had this notion [in
      mind here] I judge from the fact that he straightway drew the same
      [conclusions] concerning rest as opposed to the whole [of] motion [i.e.
      including halting].

1009,1   **239a10-22**: 'Nor is there [a time] when first a thing at rest was
          at rest' to 'for every one is partitionable to infinity.'

          [Nor is there a first time in which what is at rest was at rest;
          for a thing cannot be at rest (any more than it can move) in a
          partless now. The same conclusion follows also from the
          definition of being at rest as being in the same state both now
          and earlier.]

      After having proved that in the case of every motion, both simple
      and that [so] called in respect to halting, it is not possible to take a
      first [bit] of it, but rather that whatever one takes as first, there is
5   something earlier than this too, here he proves the same things
      concerning rest as opposed to the whole [of] motion. The cause in the
      case of rest too is the same as that in the case of motion; therefore he
      also uses the same demonstration. For if rest is a privation of
      motion, and what is so constituted as to move is deprived [of motion]
      when it is not moving in that [time] in which it is so constituted as to
      move, [then] it is necessary that it also be at rest in time, and not in
10   an atomic [i.e. indivisible] thing, just as [it is necessary that] it not
      move in an atomic thing. Furthermore, he proves in another way as
      well that being at rest is in time and not in an atomic thing,
      reminding us of what was previously said concerning rest, [namely]
      that what is similar both itself and [in] its parts now and earlier is
      said to be at rest [234b5-7], so that rest is defined by two nows, one
      now and one earlier. But there is time between every two nows.
15   Accordingly, all rest is in time. (For in fact the now and the earlier
      are parts of time [cf. 223a4-8, where however the formula is 'in time',
      not 'parts of time']). But it is not possible to take any first [bit] of the
      time in which what is at rest[86] began its rest. For time is continuous
      and divisible to infinity, and there is no first in a continuous thing.

[86] Reading *to êremoun* with CM (as suggested by Cecilia Trifogli) instead of *to
êremein* with Diels and the other MSS; cf. the parallel expression in connection with
change at 984,8-11, 20.

239a23-b4: 'Since every moving thing moves in time' to 'for it results that the locally moving thing is at rest.'

[In the exact amount of time (the first time) that a moving thing takes to reach its destination, it does not rest at any stage of its journey, because it is at a stage not for any stretch of time, but only in a partless 'now', and in a partless 'now' it does not rest.]

In this [passage], Aristotle offers, with demonstrations, the paradoxes of the theorems concerning motion which, indeed, one would not easily think up or trust. Those things too which were demonstrated immediately before were amazing, [namely] that it is impossible to take a beginning either of motion or of rest, and it is obvious that neither [is it possible to take] an end [of motion or of 25 rest] for the same reason. But still more incongruous is what is proved here, which indeed he seems to offer in this [passage] also for the sake of resolving Zeno's argument concerning motion. For since, he says, a moving thing moves in time, in the time in which something moves in its own right – that is in the [time] as a whole in which [it moves], and not by moving in some one of the [parts] of that [time] – which indeed was previously called 'in a first [time]' 1010,1 (for in the case of time 'in a first' and 'in its own right' and 'not by virtue of [being] in some one [of the parts] of that' mean the same thing for him) – in this time it is impossible 'to be at some bit' of the interval in which it is moving, in such a way that both the moving thing itself and each of its parts are in a same interval that is equal to themselves. He proves this as follows: If a thing that is moving in 5 some first time is in some same [interval], both itself and its parts, [then], since every time is divided into an earlier and a later time, both [the moving thing] itself and its parts will be in some same [interval] now and earlier. What is – both itself and its parts – in the same [interval] for some time is at rest. Consequently, a thing that is moving in some first time, if it is in some same [interval], both itself and its parts, is at rest. But in fact it is impossible for 10 something, when it is moving, to be then at rest. Consequently, something that is moving in a first time is not (both itself and its parts) in some same [interval]. That this is being at rest, he proved by the [definition]: 'For this is how we say that [something] is at rest, whenever it is true to say that in different nows both the thing itself and its parts are in the same [interval].' What was proposed has, 15 accordingly, been proved, that something that is changing in respect to a first time cannot be at something at which both it itself and its parts are changing. For this is what '[cannot be at something] as a whole' means [in Aristotle's text]. The reason is that every time is

divisible, so that it will be true to say that in different parts of it [i.e.
the time], earlier and later, both the whole and the parts [of the
20   changing thing] are in the same [interval]. But this was [defined as]
being at rest, not moving, so that something that is changing in a
first time will not be, both itself and its parts, at something. But 'if
the moving thing is in no time at any [bit] of the interval on which it
is moving, and does not occupy a place equal to itself, how does it
accomplish the whole interval in this time, and [how] will it entirely
traverse it? [We may reply] that it is at some [bit] of the interval in a
25   now; this is a limit of time' [quoted from Themistius 198,25-8]. So
that if it is said to be in time, it is not in a first [time] nor in its own
right, but rather insofar as it is in some – not portion of the time, but
rather limit. Therefore, [Aristotle] himself, when he began his
argument, said 'not by virtue of [being] in some one [of the parts] of
that [i.e. time]', having in view the limit of it, [that is] a now.[87] In a
now alone does it occupy a place equal to itself, in which [place] it is
able also to have been at rest [i.e. over a stretch of time]. However, it
30   is not at rest in it then [i.e. in the now]. For everything that is at rest
is at rest in time, not in a now, just as it also moves in time. For it is
1011,1 also at rest in that in which it moves. Therefore, it is possible to say
truthfully that something does not move in a now, but not possible
[to say] that it is at rest in a now. That 'in time it is not possible to be
at something that is at rest' was said about a moving thing,[88]
because a moving thing is at something equal to itself, in which it
5   can also be when at rest; for it is at rest in a thing equal to itself. But
in fact, a moving thing is in something equal to itself not in time, but
rather in a now. For if in time, it would be at rest; for what is itself
and its parts in something in some time was proved to be at rest.
The same thing, accordingly, will be both at rest and moving.

10   **239b5-9**: 'Zeno reasons falsely' to 'nor any other magnitude.'

[Zeno's argument concerning the arrow (that an arrow cannot
move because at each partless 'now' it is at rest) is false, for it
presupposes that time is made up of indivisible nows.]

The puzzles of the extraordinary Eleatic Zeno, about whom the
sillographer [Timon, a writer of lampoons called silloi] said: 'great

[87] Cf. 218a24; unlike a limit, a part is an extended portion of time.
[88] Ross emends *to êremoun* in Aristotle's text to *ti êremoun*, yielding 'in time it is
impossible that it be at rest at something', this is undoubtedly correct, but Simplicius
was, of course, attempting to explain the received text, out of which, says Ross, 'it is
impossible to get a good sense' (comm. ad loc.). Simplicius takes 'at' (*kata*) to mean 'in
the same place (or state) as'.

strength, indefeasible, of double- tongued Zeno', have become, for us,
responsible also for these amazing theorems of Aristotle's   15
concerning motion; for [Aristotle] offered us what was previously
said about motion on the grounds that it was useful toward the
resolution of Zeno's argument. Therefore he both straightway
mentioned the puzzle of Zeno and he produced its solution from
what was said immediately before. Zeno's argument, assuming in
advance that 'everything, when it is at something equal' to itself,
either moves[89] or is at rest, and that nothing moves in a now, and   20
that a thing that is moving locally is always at each now in
something equal to itself, seemed to syllogize as follows: an arrow
moving locally is in every now at something equal to itself, so that [it
is so] also in every time; what in a now is at something equal to itself
does not move, since nothing moves in a now; what is not moving is
at rest, since everything either moves or is at rest; consequently, the   25
arrow moving locally, while it is moving locally, is at rest at all of the
time of its motion. What could be more incredible than this? Having
said that everything either is at rest 'or moves when it is at
something equal [to itself]', he added: 'a thing moving locally is
always in a now.' Now, it is obvious that a thing moving locally is at
something equal [to itself] in a now. [Aristotle] himself added the
solution by objecting to the [proposition]: 'a thing moving locally is   30
always at something equal to itself.' For it is not the case that if in
every time there are nows, and in a now [the moving thing] is at   1012,1
something equal to itself, it is [so] already also in [a stretch of] time
and always, because of the fact that a now is not a portion of time,
but rather a limit, in which it has been proved that a moving thing is
in something equal to itself, in the theorem before this one. If,
accordingly, time is not composed of nows that are indivisible (he
added this [i.e. 'indivisible'], since we also call a certain [stretch of]
time 'now', as being present), then neither is the locally moving   5
arrow in something equal to itself during the time in which it is
moving locally, so that neither is it at rest in that time. That too has
not been assumed correctly, [namely] that everything either moves
or is at rest. For it was proved that in a now something neither
moves nor is at rest; for it is not the case that if something does not
move, it is necessary that it be at rest. Thus, from the previously
stated theorem which proves that a moving thing will not be at   10
something during any [stretch of] time, but rather [only] at a limit of
time, he resolved the [proposition] that a thing moving locally is
always at something equal to itself. For it is not [so] always, but
rather in a now, when it happens that it neither moves nor is at rest.

[89] Ross brackets *ê kineitai* in Aristotle's text, giving 'everything is at rest when it is
at something equal [to itself]'; because he retains the words, Simplicius'
reconstruction of Aristotle's argument is elaborate.

One must consider this now, in which a moving thing occupies
15   something equal to itself, as existing only potentially, just as the
point in a line. For if [the now] had actually been taken everywhere
in the time, the time would be divided into partless things and
would be composed of these, and there will be an actually infinite
number of things. Zeno's argument set the puzzle in motion on the
basis of time being composed of nows, since if at all events this
should not be granted him, it is not true that what is moving locally
is always in something equal [to itself].

20   **239b9-14**: 'The arguments concerning motion are four' to
'concerning which we decided in earlier arguments.'

[The four arguments of Zeno: (i) Moving is impossible because
to move any distance a thing must always have moved half
that distance earlier.

He says that Zeno's arguments concerning motion are four, by
means of which he exercised his audience and seemed to deny the
most self-evident thing among the things that are, [namely] motion,
so that even Diogenes the Cynic, having heard these puzzles once,
25   said nothing in reply to them, but stood up and walked and by
means of self-evidence itself resolved the paradoxes in the
arguments [cf. Diogenes Laertius 6.39]. He says that the arguments
that provide difficulties to those attempting to solve [them] are four,
either because all the arguments concerning motion were four, all of
them happening to be difficult to confront, or because, although
there were more, the most serious were four. It was fitting for one
30   who was writing a treatise concerning motion to resolve the
arguments that attempted to deny it, and above all those which
were attempted on the basis of physical principles, such as Zeno's
1013,1   arguments also were, since they exploited the cutting of magnitudes
to infinity, and the fact that everything moves or is at rest, and so
many such things.
     The first of the four arguments proving that certain impossible
things follow upon the fact that motion exists is as follows: if there is
motion, it is necessary that a moving thing will entirely traverse an
5   infinite number of things in a finite time; but this is impossible:
consequently, there is no motion. He used to prove the conditional
premise on the basis of the fact that a moving thing moves [over] a
certain interval. Given that every interval is divisible to infinity, it
is necessary that the moving thing first traverse half of the interval
[over] which it is moving, and then the whole: but, even before [it
traverses] half of the whole, [it must traverse] half of that, and again

half of this. If, accordingly, the halves are infinite because it is   10
possible to take a half of everything that is taken, and it is
impossible to traverse an infinite number of things in a finite time –
this Zeno used to take as self-evident :Aristotle had this argument
in mind earlier when he said that it is impossible to traverse an
infinite number of things and to touch an infinite number of things
in a finite [time, cf. 233a26-27]; but in fact every magnitude has   15
infinite divisions; consequently, it is impossible to traverse any
magnitude in a finite time. [Aristotle] resolved it by saying that
there is not an infinite number of things in the interval actually, but
rather potentially, and that nothing can entirely traverse in a finite
time halves insofar as they are actually infinite, but [insofar as they
are] potentially [so] nothing prevents [this]. For neither will the
moving thing traverse the interval by dividing it into infinite halves,   20
but rather as as single and continuous thing. Therefore it is not
prevented from entirely traversing things infinite in this way, for if
it should divide [the interval] into halves, it will no longer move as
though on a single and continuous thing, nor will the motion be
continuous. Thus, accordingly, he resolved the argument according
to a distinction [in the kinds] of infinity, and further also on the
basis of the fact that the puzzle[90] is alike both in the time and in the
interval, so that [the moving thing] will traverse an infinite
[interval] not in a finite time, but rather in a similarly infinite   25
[time]. For in fact the time too is divisible to infinity. But the infinity
is potential and not actual. Accordingly, it will traverse a potentially
infinite number of [bits] of the magnitude in a potentially infinite
number of [bits] of the time.

239b14-30: 'The one called "Achilles" is the second' to 'these,
accordingly, are two [of the] arguments.'   30

[(ii) The paradox of Achilles: that a faster thing will never
overtake a slower, since the slower has always advanced by the
time the faster has reached its starting point.]

This argument too has been attempted on the basis of division to
infinity in another version. It would be as follows: if there is motion,   1014,1
the slowest will never be overtaken by the fastest. But in fact this is
impossible. Consequently, there is no motion. The conditional
premise is self-evident, and he establishes the additional premise
which says that it is impossible for the slowest to be overtaken by   5
the fastest, by taking a tortoise as the slowest, which the story too

---

[90] The Aldine reading *apeirian*, 'infinity', instead of *aporian*, 'puzzle', is attractive.

took as slow by nature in the contest with the horse [cf. Themistius 185,6], and Achilles as the fastest, who seemed so much the swiftest of foot that 'help-foot' [*podarkês*] seems to be his personal epithet in Homer because the speed of his feet helped [*arkein*] both himself

10   and his allies. The argument was called 'Achilles', accordingly, from the fact that Achilles was taken [as a character] in it, and the argument says it is impossible for him to overtake the tortoise when pursuing it. For in fact it is necessary that what is to overtake [something], before overtaking [it], first reach the limit from which what is fleeing set forth. In [the time in] which what is pursuing arrives at this, what is fleeing will advance a certain interval, even if it is less than that which what is pursuing advanced, by virtue of the

15   fact that it is slower. Nevertheless, it will advance: for it is not, indeed, at rest. And in the time again in which what is pursuing will traverse this [interval] which what is fleeing advanced, in this time again what is fleeing will traverse some amount by so much less than that which it moved earlier as it itself is slower than what is pursuing. And thus in every time in which what is pursuing will traverse the [interval] which what is fleeing, being slower, has [already] advanced, what is fleeing will also advance some amount.

20   For even if it is always less, nevertheless it will itself too traverse some amount if it is moving at all. By taking one interval less than another to infinity by the cutting of the magnitudes to infinity, not only will Hector not be overtaken by Achilles, but not even the tortoise [will be]. For, let it be hypothesized that the prescribed [interval] is a stade, and that the tortoise is ahead [by] half a stade, and that Achilles moves [a distance] ten times greater than the

25   tortoise in the same time. Once Achilles, then, has begun to pursue the tortoise from the beginning of the stade, in however much [time] he will advance the half-stade, so as to reach the half from which the tortoise set forth, the tortoise too will be traversing a tenth of the remaining half-stade. Again, in however much [time] Achilles will

30   traverse a tenth of this half-stade, the tortoise traverses a tenth of a tenth of a half-stade, and if upon every tenth that is taken of any

1015,1 interval it too has a tenth, the tortoise will forever be somewhat ahead of Achilles and never will either of them entirely traverse the stade.

This argument too, accordingly, is of this sort. [Aristotle] says that it is the same as the one before this, because the plausibility of this

5   one too is dependent upon the division of a magnitude to infinity. It seems to differ from the former by not always dividing according to a dichotomy and in half, but rather according to some other ratio according to which the motion of the swiftest exceeds that of the slowest, whether the ratio be ten times or some other. That not even the fastest pursuing the slowest will overtake it has been added in

tragic style to this argument, with both Achilles and the tortoise   10
brought in as characters as though in a tragedy or comedy. The
solution here too is the same, from the distinction [in the kinds] of
division to infinity: for there is not an actually infinite number of
things on a continuous thing and a continuous motion on it. For they
do not still remain continuous when divided, for insofar as it [i.e. the
continuous thing] is divided, it is separated[91] and does not keep the
magnitude continuous. The slower thing, when it has a lead, is not   15
overtaken, but it is, nevertheless, overtaken. For if the argument
grants that [the faster thing] will entirely traverse a finite [distance]
such as a half-stade and a tenth of a half-stade, it is obvious that it
will traverse the entire stade as well, and faster than the slower
thing that is moving in the same time, and will attain what, indeed,
the slower thing traversed, and in this way will overtake it. For the
faster, having traversed in one hour what the slower moves in ten   20
hours, will arrive at the same thing as the slower. Here too the
puzzle has arisen through assuming an actually infinite number of
things, which indeed is not true and above all when magnitude and
motion are kept continuous. For even if it is necessary that a moving
thing move some first part and in this way [move] the whole
[interval], it is nevertheless not necessary that this [part] have been   25
marked off: rather, it is necessary that it not have been marked off,
so that the continuous thing not be divided.

239b30-33: 'The [argument] mentioned just now is the third' to
'there will be no syllogism.'

[(iii) A flying arrow cannot move.]

Third, he says, is the argument of Zeno which he solved shortly
before, that says that an arrow moving locally halts in its moving,
since it is necessary that everything either move or be at rest, and a   30
thing that is moving locally is always at something equal to itself.
What is always at something equal to itself does not move:
consequently, it is at rest. This argument assumes a falsity,
[namely], that it is at something equal to itself in time. For it has   1016,1
been proved that what is moving locally is at something equal to
itself in a now only. This [Zeno] too admits, but he does not consider
that in a now nothing either is at rest or moves. Since the nows do
not fail [throughout the stretch of time], he believes on this account
that it [i.e. the arrow] stands still in the time of its motion,   5

---

[91] Reading *diistatai* with CM, instead of *te histatai* (= 'it halts') with Diels and A,
which is difficult to construe if the subject is 'a continuous thing'.

syllogizing nothing [validly] on the basis of this, since nows are not continuous with each other nor does time arise out of nows.

> **239b33-240a18**: 'The [argument] concerning things [moving] in a stadium is fourth' to 'the falsity results through what has been said.'

> [(iv) The stadium: with two rows of objects moving in opposite directions, and a third row stationary, it appears that half an interval of time equals double the interval. The fallacy lies in supposing that it takes the same amount of time to pass a stationary and a moving body. Thus, if A's are stationary, while B's and C's move in opposite directions, the B's will pass all the C's but only half the A's in the same amount of time.]

The fourth of Zeno's arguments concerning motion, this one too
10   reducing to impossibility the fact that there is motion, was something like this: if there is motion, of magnitudes that are equal and of equal speed one will move [with] double the motion of the other, and not [with] an equal [motion], in the same time. This is absurd, and what follows upon this is also absurd, [namely] that the same and equal time is simultaneously double and half. He proves it by taking it as agreed that things of equal speed and equal [in
15   magnitude] have moved an equal interval in an equal time: and further, indeed, of things of equal velocity and equal [in magnitude], if one has moved half, and the other double [the interval], the one has moved the half [interval] in half the time, and the other the double in double the time. These things having been assumed in
20   advance, a stadium was hypothesized, for example DE, and four magnitudes, or however many, only even [in number], so that being of equal volume (and as Eudemus [fr. 106 Wehrli] says, cubes) they
25   have a half, on which A's [are labelled], such that these, standing still, hold the middle interval of the stadium. Of the ones that stand still, [Aristotle] defines as first that near the beginning of the
1017,1   stadium, which is at D, and as last that near E, and he takes another four volumes or cubes equal to the ones that are standing still both in magnitude and in number, on which B's [are labelled], beginning from the beginning of the stadium, and ending at the middle of the four A's, these [i.e. the B's] moving toward the
5   extreme, E, of the stadium. Therefore he also calls 'first' the [B cube] at the middle of the A's, since it is in front of the rest in the motion toward E. He took the volumes as even [in number] for this [very reason, i.e.] that they would have a half. For he needs this, as we shall learn. Therefore he also puts the first B against the middle of

the standing A's, and then he takes still other [volumes] equal in
magnitude and number to the B's (it is obvious that [they are equal]
to the A's as well), on which C's [are labelled], moving opposite to the   10
B's. For while the B's are moving from middle of the stadium, in
which there was also the middle of the A's, toward the extreme, E, of
the stadium, the C's move from the extreme part, in which E is,
toward D which is in the beginning of the stadium, and the first of
the four C's, obviously, is the one facing toward D, toward which the
C's have their motion. He puts the first C against the first B. This   15
being the position hypothesized at the beginning, if, while the A's
are standing still, the B's move from the middle of both the A's and
the stadium toward the end, E, of the stadium, and the C's [move]
'from the extreme' [i.e.] of the stadium toward the beginning,
obviously (for it is not, indeed, 'from the extreme B', having
discovered which, as it seems, in certain manuscripts, Alexander   20
was constrained to say that what [Aristotle] earlier called the 'first
B' he here called the 'extreme' [or 'last']), 'it results that the first B
and the first C are simultaneously at the extreme' [i.e.] of their
motion,[92] since they 'move past each other' and at the same speed or
else [they are simultaneously] 'at the extreme' [i.e.] of each other.
For if the first C is against the first B in the beginning, and they
move oppositely at equal speed and entirely traverse one another,
the first B will be at the extreme [or last] C, and the first C at the   25
extreme B. And this would be the result that 'the first B and the first
C are simultaneously at the extreme' of the things that 'move past
each other'.[93] For the motion past each other produces their coming
to be in the extremes of each other.

[Diels' diagram]

		A	A	A	A	
B	B	B	B	→		
		←	C	C	C	C

[N.B. Diels does not reproduce the diagrams in the MSS]

He says next 'it results that C' too (the first one, obviously) 'has
traversed past all the A's, while B [has traversed] past half the   1018,1
A's.'[94] It is obvious that B, beginning from the middle of the A's,

---

[92] Omitting Diels' full stop after *einai*, clearly a typographical error.

[93] The difference in the explanation depends on two possible interpretations of the
genitive, *kinoumenôn*, in Aristotle at 240a10.

[94] For 'all the A's', a few MSS of Aristotle read 'all the B's' (printed in brackets by
Ross); for 'half the A's' the MSS of Aristotle read 'half of them'.

moved through two A's (or through half of them, however many even ones there are), in as much [time] as C will traverse through double the B's. For the first B made a beginning from the middle of the A's.

5   And in as much [time] as B moves past the two extreme A's that are standing still, the first C moving opposite to the B's will traverse through four B's. For the two motions of the things moving opposite accomplish double the interval of the one [motion with] which B moves past the standing A's. This too is obvious. But how has C

10   traversed 'past all the A's'? For it was not moving past these, but rather past the B's, nor was it moving from the beginning of the A's, but rather from the beginning of the B's, which was at the middle of the A's. [We may reply] that the B's were also equal to the A's. Accordingly, in as much [time] as C has moved past the B's, it would have moved past the A's as well, which are equal to the B's. And the fallacy here is that he took what is moving past equal things as

15   simply moving in an equal time, without reckoning in addition that of the equal things some were moving opposite while others were standing still. Having assumed nevertheless that the C's will traverse both the B's and the A's in an equal time, since in as much time as the first B will traverse the two A's, the C [will traverse] the four B's or rather the four A's, he inferred that B, although it is equal in speed to C, in the same time moves half of what C moves, which indeed is contrary to what has been agreed upon previously

20   and to what is self-evident. For things of equal speed in an equal time move [over] an equal [interval], but [only] when they are similar, such that either both move past things that are standing still, or both past things that are moving, and not when some [move] past things that are standing still, as B, and others past things that are moving opposite, as C. Further, the time too, in which B moves through the two A's, is half of the time in which C moves through the

25   four B's, since the A's are equal to the B's, and both B and C are of equal speed. The time in which B was moving through the two A's, and [that] in which C [was moving through] the four B's, seemed also to be equal or rather the same. It will result, accordingly, that the same magnitude too is both double and half, since in the same

1019,1 time of things of equal speed, one, B was traversing two A's, while another, C, [was traversing] four B's, with the B's being equal to the A's and that the same time is both double and half, since the time, in which B was traversing through the two A's, was both half of the time in which C [was traversing] through the four B's, and the same. The [statement] 'for either is equal alongside each' signifies that

5   both B and C, being of equal speed, spend an equal time alongside each of the things through which they move, both the B's and the A's. But if an equal [time], [then] it is obvious that the time in which C will traverse the four B's is double, while that in which B [will

traverse] the two A's is half, or rather that the [time] in which C will
traverse the four A's [is double] that in which B, being of the same
speed as it, [will traverse] the two A's. For it has been said that in
the [time] in which C will traverse the B's, it will also [traverse] the
A's.

Having said that 'it results that C has traversed past all the A's',   10
because it traverses past all the B's, and meanwhile having posited
the absurdities that follow, [namely] that half the interval is the
same as double the interval and that half the time [is the same] as
double the time, he added that 'it results simultaneously that the
B's', too, 'have passed by all the C's', just as the C's have [passed] by
the B's. 'For the first C and the first B will be simultaneously at the   15
contrary extremes', C at the beginning of the stadium, and B at the
end, which indeed are both extremes. For, moving opposite [each
other] at the same speed and being equal, and having begun from
the beginning of one another, they simultaneously overtake the
limits of one another, C 'being for an equal time alongside each of
the B's',[95] [just] as much [time] as alongside each of the A's, even if C
does not move past the A's, because 'both' the C's and the B's, being   20
equal and of equal speed, 'are for an equal time alongside the A's'.
By this [statement], [Aristotle] showed why [Zeno] took the C's to
have moved past the A's even though they have not moved [by them,
but rather by the B's], [namely] that it is on account of their equality
and equal speed in respect to the B's at the same time, too, by having
added the [phrase], 'as he [i.e. Zeno] says',[96] [Aristotle] showed that   25
the fallacy has arisen [precisely] with this. For C is not alongside
each of the things moving opposite and [each] of the A's that are
standing still for an equal time.

Alexander says that the [statement] 'being for an equal time
alongside each of the B's, [just] as much [time], indeed, [as alongside
each] of the A's', can come next after the [statement], 'it results that   30
C has traversed past all the A's', and then, next, the [statement]
'while B [has traversed] past the halves...'[97] to '...for the first C and
the first B will be simultaneously at the contrary extremes', and
next after these [words] the [statement] 'because both are for an
equal time at A'.[98]   1020,1

Such, accordingly, is the argument, and it is most silly, as
Eudemus [fr. 106b Wehrli] says, because it has the fallacy
conspicuous, since [Zeno] rules that things that are equal and of

[95] Ross, in his text of Aristotle, brackets the words cited by Simplicius.
[96] Ross, in his text of Aristotle, brackets these words.
[97] Cited by Simplicius at 1018,1-2 as 'while B [has traversed] past half the A's', i.e.
ta hêmisê A instead of *ta hêmisê*; Alexander's version here agrees with the MSS of
Aristotle.
[98] Cited by Simplicius at 1019,21 as 'for an equal time alongside the A's'; Simplicius'
version here agrees with the MSS of Aristotle.

equal speed move [over] an equal interval in an equal time if one moves past a moving thing and the other past a thing at rest, which indeed is self-evidently absurd. For things of equal speed moving
5   opposite each other withdraw double the distance in the same time in which what is moving past a thing at rest is separated [by] a half, even if it is equal in speed to the former things.

240a19-29 'Nor, indeed, in respect to change in a contradictory pair' to 'as a whole [it will be always] in neither.'

[Other puzzles involving change (i) That a thing changing from not-white to white is neither white nor not-white. But a thing not wholly in white (or not-white) may be called white (or not-white) if the more numerous or most important parts of it are in white (or not-white.]

Having resolved the arguments of Zeno that seem to deny motion,
10   he brings in other arguments as well which are posed sophistically to the same [end], so that he may resolve their sophistries as well. The first of these, which attempts to deny motion on the basis of change according to contradictory pairs, is as follows since it is posited that a thing that is moving and changing from something to something is in neither when it is moving, neither in that from
15   which (for this [i.e. what is in that from which] is not yet moving); nor in that to which (for this is no longer moving),[99] as Aristotle himself also teaches in this book [cf. 234b10-13], but there is change also according to contradictory pairs, what is changing according to contradictory pairs will be in neither portion of the contradictory pair, which indeed is absurd, for [then] there will be something in between the contradictory pair. But, if this [latter] is not [the case], [then] it is not true that what is changing is neither in that from which it is changing nor in that to which it is changing, in the time
20   in which it is changing. For in fact a change according to a contradictory pair has a 'from which' and a 'to which', for it changes from white to not white and from not white to white and from being to not being and from not being to being. If, accordingly, it is necessary, when there is a change, for what is changing to be in neither [portion] of the contradictory pair, and [if] this is impossible, because nothing escapes the antithetical opposition of a contra-
25   dictory pair, [then] there could not be change. Solving this argument, [Aristotle] says that the thing changing according to a

---

[99] Reading *touto* bis with the MSS (cf. 1034,23-4), instead of Diels' emendation *tote*, which gives: 'for then it is not yet/no longer moving'.

contradictory pair is as a whole in neither portion of the contradictory pair, but some part [of it] is in each. And this has been proved earlier, when he said that of something that is changing it is necessary that some of it be in that from which, and some in that to which [cf. 234b15-17]. If this [is the case], the changing thing as a 1021,1 whole will neither be in one or the other part of the contradictory pair nor will there be anything in between the contradictory pair, in which it will have to be, but rather it is in both [parts of the contradictory pair]. What, accordingly, will it be called: white or not white? For it must be one or the other by the axiom of contradiction, and whether it is white or not white, it is not in both. Resolving this 5 objection, he says that it is not necessary that it be in either as a whole, for it to be called white or not white. For in fact not only what is such as a whole is called white, but also what has 'most [of its parts] or its most important parts' such. Similarly, what does not have most [of its parts] or its most important [parts] white is not white, so that it will be called [white or not white] on the basis of 10 that in which the greater or the more important [part] is. But as a whole it will be in neither in [the course of] its changing. How, accordingly, is the same thing white and not white?[100] [We may reply that] first this distinction too, [namely] that it is not at both things [i.e. white and not white], was invoked on the strength of the [axiom of] contradiction, but then [the further distinction was invoked] that while 'it is simply [or absolutely (*haplôs*)] not in white' cannot be true together with 'it is in white', 'it is not in white as a 15 whole' can be true together with 'it is partly white'. What is not white as a whole is also called white, since 'it is simply not white' and 'it is not white as a whole' are not the same thing. Having taken change according to a contradictory pair in the case of white and not white, and having solved the puzzle in this case through the addition of 'as a whole' (for it is not true that what is changing is simply in neither [part] of the contradictory pair, as the poser of the 20 puzzle ruled, but it is true that it is not [in either part] as a whole) for 'it is not as a whole' and 'it is not' are not the same thing; – having solved it, accordingly, in this way in the case of white and not white, he added that 'it is similar also in the case of being and not being', in the case of which it is more proper to see only a change according to contradictory pairs, for white properly changes also into its contrary, black, or into what is between, while what is 25 [changes] only into what is not and what is not into what is. Having added, accordingly, both being and not being, he [then] added in general that the puzzle is to be solved in this way in respect to every

[100] Punctuating with a query after *leukon*, instead of a comma (as suggested by Alain Segonds); *ê* here (like *mêpote*) introduces a rejoinder to a puzzle or question, a common idiom in the commentators on Aristotle.

contradictory pair. For what is changing will, like all other things as
well, be 'necessarily in one or the other' part of the contradictory
pair, but not always as a whole. For in [the course of] changing it is
not [in one or the other part] as a whole. Therefore, when the
30   contradictory pair is stated simply, the solution to the puzzle is that
which distinguishes 'it is not as a whole' from 'it is not'. If, with the
addition, 'as a whole', the premise should be stated, 'it is white as a
1022,1   whole', [then] the negation that has 'not as a whole' is true in these
cases. But it is obvious that it is possible to say that both what is
coming to be and what is perishing are in one way and are not in
another way, in respect to an approach toward being and not being.
What are the 'other things' in respect to a contradictory pair? [We
may reply)] All things that are, whenever they are divided by an
affirmation and a negation that is opposed contradictorily [e.g. 'is
white' and 'is not white'].

5   **240a29-b7)** 'Again in the case of a circle and a sphere' to 'and
the other things that move in themselves.'[101]

[Second puzzle: (ii) A rotating object is in the same place over
time and hence is at rest. But its parts are not in the same
place, nor indeed is the whole) for circumferences beginning at
distinct points A, B, or C are not the same, except incidentally
(as are a musical man and a man).]

That he is still objecting to some argument of those who deny motion
is obvious from the fact that the argument before this also, since it
followed upon what was said against Zeno, began in this way) 'nor
10   indeed in the case of change in a contradictory pair will there be any
impossibility for us' with regard [that is,] to the fact that there is
motion. It is obvious, accordingly, that here too, as though speaking
about the same thing, he says 'again in the case of the circle and the
sphere and things that move in themselves, that it will result that
they are at rest', as the quibblers think, does not, this either, entail
any impossibility [consequent] upon there being motion. Aspasius
too says that this argument also has this aim, and before him
15   Eudemus [fr. 107 Wehrli]. Alexander, however, believes that what is
said here is said against those who deny the definition of rest
expounded by him [i.e. Aristotle]. For since he said that what is,
both itself and its parts, in the same thing for a certain time is at
rest, he proves that this is not denied by virtue of things that move
in the same place, as some think) such things are those that rotate,

---

[101] Reading *hautois* instead of Diels' misprint *autois*.

which seem to be in the same place for a certain time, both   20
themselves and their parts, although they are moving. 'He proves,
accordingly,' says [Alexander], 'that these things do not come under
the definition of rest.' That he proves this is obvious) but he proves it
not by confirming the definition of rest, but rather [by confirming]
that motion is not denied on the basis of this puzzle either. The
puzzle seemed to be something of this sort) if the things that seem to   25
be moving [with] the first and most important motion, [namely],
circular [motion], are shown not to be moving but rather to be at
rest, there would not be motion. The conditional premise being   1023,1
readily acceptable (for who does not agree that if the first and most
important of motions is not motion but rather rest, there would not
be motion), they seemed to demonstrate the additional premise by
taking it as agreed also on the basis of what has been proved that
things that are, both themselves and their parts, in the same place   5
for a certain time are at rest, assuming in addition that things
moving circularly or spherically are both themselves and their parts
in the same place for a certain time, and inferring by means of these
[premises] that things that seem to move circularly or spherically
are at rest. [Aristotle], then, solves this argument too by proving
that things moving in this way are neither themselves nor their
parts in the same place for any [stretch of] time. First he says that   10
the parts are not in the same [place] for any time, for each of the
portions turns up at different times in different [places] and at a
different [part] of the surrounding [medium]) for if the parts do not
always maintain the same position, it is obvious that each is at
different times in a different place. Now, a thing that is – both itself
and its parts – similar for a certain time, was [said to be] at rest. But   15
not even the whole remains in the same [place]) for if all the portions
[move], the whole moves as well according to its portions. [Aristotle]
himself proves further, for good measure, that the whole is moving
not from the parts, but rather in its own right as a whole. For he
says that the whole too is always changing to another thing, [i.e.]
place, obviously, which indeed seems amazing, for if a [circumfer-
ence] taken from A to A is one circumference, and one from B to B is
another, and so [on] at each point, then the local circumference [i.e.   20
the inner boundary of the surrounding medium, cf. 212a20-21] too
has been similarly divided into [circumferences] corresponding to
these. Also, each is a whole circumference both of what is in a place
[i.e. the rotating thing] and of the place [i.e. the unmoving inner
limit of the surrounding medium]. Since the points [on the rotating
thing] move [Simplicius neglects to say, 'incidentally' (cf. 240b8-9)],
the whole is always changing to the place of a different whole, since   25
the circumference from A to A is not the same as that from B to B
either in the place or in what is in place, except incidentally, because

all [the circumferences] are of [i.e. belong to] the same circumference, just as 'a musical man' [and a man[102] are the same, but incidentally], 'because it is incidental to' [e.g.] Aristoxenus [a famous musical theorist and pupil of Aristotle], who is a man, also to be musical [i.e. being musical (incidentally) and being a man are of or belong to Aristoxenus; cf. *Metaphysics* 1017a9-18]. Thus one circumference would always be changing into another, for they take up in turn one another's positions. For where the circumference
30   from A to A was now, there shortly after will be that from B to B. Since they change, they [i.e. the parts or individual circumferences][103] will not be in the same [place] for a certain time nor will the whole. Nor, consequently, will it be at rest. Having proved in the case of the circle that neither the parts nor the whole are at rest (for
1024,1 [he proved that] they do not remain in the same [place]), he says that there will be the same argument 'both in the case of a sphere and in the case of other things that move in themselves' and that do not depart from their own place, such as both tops are and anything else that spins round.
5   It should be known that whereas Aristotle said that the circumference first[104] taken from A was not the same as that [taken] from B, Eudemus, lover of truth, says [fr. 108 Wehrli]) 'it takes investigation whether it should be said that the whole [circumference], beginning[105] from anywhere whatsoever, is the same circumference [as these] or another.'

240b8-[8D] 241a6) 'These things having been demonstrated' to
10   'just as if [he were to make] time out of nows or a length out of points.'

[A partless thing can move only incidentally by being in a moving body, like a man in a boat or the parts of a rotating sphere. First proof: if a thing is changing from AB to BC in the time D, it must be wholly in AB or BC (in which case it is not changing), or partly in each (in which case it is not partless).

---

[102] In quoting Aristotle, Simplicius omits the words *kai anthrôpos*, 'is also a man' (240b4).

[103] So the Aldine, which supplies *ta merê* before *oude to holon*; Diels, in the apparatus criticus, would supply *sêmeia* ('points') with *tina* (line 31), and take that as the subject of the sentence, which is impossible; *tina* must, here as elsewhere, modify *khronon* ('time'). Alternatively, one might emend *metaballonta* to *metaballon*, yielding 'since it changes, not even the whole will be in the same place for a certain time'.

[104] Reading *prôton* after *periphereian* with the MSS, and omitting *prôtou* before *A*, which appears only in MS A (so Wehrli).

[105] Reading *arxamenên* with most MSS and Wehrli, instead of *arxamenôn* with A1 and Diels.

For a partless thing to move, all continua (e.g. time, length and
motion) would have to be composed of partless things.]

Having proved earlier that nothing either moves or is at rest in a
partless thing, but rather that while it is true to say that it does not
move in a partless thing and that it is at something in a partless
thing, it is not, however, at rest, and on this account having
mentioned Zeno's argument as fallacious, when he rules that
everything either is at rest or moves when it is at something equal
[to itself], and having resolved the argument that tried to deny   15
motion on this basis, and then having brought in the other
[arguments] of Zeno as well that try sophistically to deny motion
and having resolved them, and having mentioned two other
arguments as well that have the same aim and having resolved
those too, the things that were said about motion immediately
before and those earlier about the fact that something neither moves
nor is at rest in a partless thing 'having been demonstrated', he   20   ˙
says, 'let us say' next after these things 'that a partless thing cannot
move except incidentally', by being in a moving body or magnitude.
How this is meant he indicated in the example in respect to the boat:
for things in it that are not moving in their own right themselves too
move locally incidentally, by virtue of being in the boat that is
moving. To [the word] 'body' [Aristotle] added [the word]   25
'magnitude', as Alexander says, since he is calling a line and a
surface 'magnitude', and these, indeed, are not bodies nor separable
from body. Aspasius says that these things are said redundantly, for
no other magnitude moves in its own right except body) and in fact   1025,1
both a line and a surface move incidentally by being in a moving
body. But [Aristotle], having started to say '[partless things move,]
for example, when a body is moving, by inhering', remarked that a
partless thing is not in a body first, but rather in a line: he therefore
added 'or when a magnitude [is moving]', by inhering [that is,] in the   5
moving magnitude of the line, which itself too is moving incidentally
by inhering in a surface, which itself too moves incidentally by
inhering in what is moving in its own right, [namely] a body. He
indicated all these things, accordingly, by adding the [phrase], 'or
when a magnitude'. Having said that what is at rest in a boat moves
incidentally by virtue of being in the moving boat, he added 'or the
part by the motion of the whole'. For in fact the part too moves   10
incidentally, whenever the whole moves, by virtue of being in the
moving whole. Intending to prove that a partless thing cannot move
in its own right, he first defines a partless thing [by saying] that it is
'what is indivisible in respect to quantity'. And having inserted this
definition in between, following upon 'or the part by the motion of   15
the whole', he adds: 'for in fact the motions of the parts are different

in respect to the parts themselves and in respect to the motion of the whole', resolving some objection bearing upon the fact that parts move incidentally. For when he was proving that motion was divisible not only in respect to time, but also in respect to the
20 portions of the moving thing, then a motion of the portions themselves in their own right too was assumed [cf. 234b21-27], but here he says that the portions move incidentally by inhering in the moving whole. He says, accordingly, that the motions of the parts are of two kinds, some being of the parts themselves moving in their own right, and others 'in respect to the motion of the whole'. He proves simultaneously, through this, also the difference between the
25 motion of the parts in a whole and that of a partless thing in a magnitude. For since he said that a partless thing cannot move except incidentally by inhering in a moving body, just as a part too is said to move by the motion of the whole, he added that the motion of a part is of two kinds. In this way, accordingly, it [i.e. a part] differs from the partless thing, because a partless thing moves only
30 incidentally, while a part [moves] both in its own right and incidentally. That there are also motions of each of the parts in their own right in the motion of the whole he proves self-evidently in the case of a rotating sphere. For the speed of portions of it toward the
1026,1 centre and of [portions] equal to these toward the outer surface are different. For unless the portions toward the outer surface[106] were much faster than those toward the centre, they would not entirely traverse so much greater an interval in the same time than the rotation of the sphere. For the things rotating on the largest circle
5 and those on the smallest come back around together. Alexander says: 'it is possible that the [phrase] "of the things toward the centre and those outside" is speaking [respectively] about the great circles drawn through the centre, such as the equinoctial and the zodiacal, and those around the poles [i.e. latitudinally]: for the latter are outside the centre of the sphere [i.e. they do not bisect it]. And obviously in the case of these [circles], conversely, those toward the
10 centre, being larger, are faster, while those outside are slower, since the points in them come back around together in the same time.' In this way, accordingly, the motions of the portions differ from one another, and differ from the whole [motion with] which the whole sphere moves, since that [motion] will traverse all the intervals in the sphere in the same time, while each of the parts [will traverse] that in respect to itself: and the motion of no part is the same as the
15 whole [motion], so that the portions too move [with] a certain motion in their own right alongside the [motion] of the whole, and the

---

[106] Supplying the iota subscript to *epiphaneiâi*, which is accidentally omitted in Diels' text.

motion of these is not only incidental, like that of the partless things.

Having proved these things concerning parts, he passes next to the partless thing and says that this 'can move' only incidentally 'like a man sitting in a boat when the boat is running, but cannot   20
[move] in its own right'. And he proves this in the case of all types of change, saying that what increases and decreases and what changes in respect to place [both] change in respect to magnitude: for place too is a kind of magnitude. Therefore he did not mention place since it is comprised in magnitude. 'From type to type' refers to change according to quality, for type [*eidos*] is a quality, while 'according to   25
a contradictory pair' [refers to] coming to be and perishing, for these [changes] are from not being to being and from being to not being. He proves it by a reduction to impossibility, using in addition what has been proved earlier concerning the fact that every moving thing is divisible [cf. 234b10]. Therefore there too he seemed to make the demonstration general for all moving things and not for things only   30
that change in respect to place. Having assumed that what changes changes from somewhere to somewhere and in time, he syllogizes potentially [i.e. according to the second figure of inference (cf. note 6)] as follows: if a partless thing changes in its own right, and [if] it is necessary that what changes be in [the course of] changing either   1027,1
as a whole in that from which it is changing or as a whole in that to which it is changing or some of it in this and some in that (for he omitted [its being] in neither as self-evidently absurd), and [if] it is impossible that a partless thing be in any of these, [then] it would be impossible for it to change in its own right. He left out the conditional premise as being self-evident from the [logical] division which was also stated earlier [cf. 234b10-17], while he proves the additional premise for each [case individually]. And first he proved   5
that there would not be 'some of it in this and some' in that from the fact that such a thing is partitionable. Then [he proved] that neither is it in that to which it is changing, for what is in that has changed, but this was hypothesized as changing. And [as for] what is left, [he proved] that neither is it in that from which it is changing 'at the time during which it is changing': for when it is in that from which it   10
is changing it is at rest and is not changing. 'For being in the same thing for some time' was said to be 'being at rest'. Thus a partless thing cannot move or change at all in its own right.

Having proved in this way that it is impossible that there be motion of a partless thing, he adds a more exact observation, saying that 'its motion would exist only in this way,[107] [namely] if time

---

[107] Simplicius here substitutes *monôs* for Aristotle's *monakhôs* (240b31), though at 1029,16, 17 he retains *monakhôs*.

15   were made of nows'. Then too its motion would be not as of
      something that moves but rather as of something that has moved,
      for each thing has moved in a now, [but] does not move. For if it were
      moving, it would be necessary that some of it be in one thing, and
      some of it in another, as has been proved earlier [cf. 234b15-16] and
      such a thing is not partless. If time will be composed of nows, it
20   follows that motion too is composed 'of moves', which indeed are
      partless limits of motion. For if the partless thing has moved at each
      of the nows, and time is made of nows, the motion too will be
      composed of just so many moves and limits of motion as the time in
      which the partless thing is said to have moved is of nows. But it has
      been proved through many [arguments] that it is impossible for any
25   continuous thing to be composed of partless [entities]. For neither
      time nor magnitude nor motion can be composed of partless
      [entities], so that not in this way either would a partless thing, as
      something that has moved in time, have motion. For in fact this is
      impossible in another way as well, for it has been proved that it is
      necessary that what has moved be moving earlier, just as [it is
      necessary] that what is moving also have moved earlier. But if there
30   was motion in a now, [then] there occurred something that had
      moved which did not go through moving toward this [state]. For
      there was not [supposed to be] moving in a now, but only having
      moved, for a limit [of moving] is in a limit [of time].

1028,1   **241a6-14:** 'Further, it is evident from these things too' to 'for an
          indivisible to move is also impossible.'

          [Second proof: a moving thing moves an amount less than or
          equal to itself before it moves a greater amount: but a partless
          thing cannot move less than itself. If it moves an equal
          amount, then a point will :impossibly; measure off a line.]

      He proves also in another way that a partless thing cannot move.
      This proof proves that it does not move in respect to locomotion, and
      it is produced upon these things being agreed in advance, [namely]
5   that a moving thing will entirely traverse either some greater local
      interval than its own volume or an equal or a lesser. For it is not
      impossible that it has moved [over] an interval less than itself. For a
      cubit-long thing can have moved [over] a half-cubit interval. For it is
      necessary that it always occupy an [interval] equal to itself, but that
      it move now [over] an equal [interval], now [over] a greater, and now
10   [over] a lesser. For whenever it leaves behind a place that is equal to
      itself in which it alone was as a whole, and takes over a [place]
      continuous with the former again equal to itself, then it has moved

[over] an [interval] equal to itself. If it should not, as a whole, leave behind a place equal to itself, it moves [over an interval] less than itself. If it should leave behind not only a [place] equal to itself, but also another [place] continuous with the former, then it moves [over] an interval greater than itself. It is necessary that this have been assumed in advance and that it is impossible for a moving thing to   15
traverse some local interval greater than itself before it should traverse one equal to itself or less. With these things, accordingly, having been assumed in advance, it is obvious that a point too will always move [over] an interval either equal to itself or less than itself before having moved [over] one greater than itself. Since it is impossible [to have moved over] a lesser 'for nothing is less than a partless thing', it is obvious that a point always moves a distance equal to itself, if indeed it moves, and [does so] even if it moves a   20
greater [distance], since it moves a greater [distance] that is composed of many equal ones. If it will always move according to an [interval] equal to itself when it entirely traverses a line, it will have measured it off. If this [is the case], the line will be composed of as many partless points as the number of times the partless thing moving on it was at [intervals] equal to itself. For everything that is   25
measured off by something has some portion of itself that measures it off, and is as many times greater than it [i.e. the portion] as the number of times it is measured off by it. But it is impossible for a line to be composed of partless things. Consequently, it is impossible that a partless thing move. And the whole composition of the argument which itself is also by reduction to impossibility is as follows: If a point moves according to locomotion, [then] moving always according to an [interval] equal to itself it will measure off the line, and the line will be made of points: but in fact this is   30
impossible: consequently, the point does not move.

241a15-26: 'Further, if everything moves in time' to 'for moving   1029,1
in a now and that an indivisible thing moves belong to the
same argument.'

[Third proof: nothing moves in a now, and the time in which a
thing moves is always divisible: in less time than it takes to
move an amount equal to itself, a partless thing must move an
amount less than itself.]

This is a third argument proving that it is impossible for a partless thing to move. And this one too will proceed as in the the case of points, but the proof fits in the case of every change, since it assumes   5
in advance as previously having been proved that every moving

thing moves in time, but that nothing moves in a now, and since this same [proof] also reduces to an impossibility, [namely] that an indivisible thing is divisible, if a partless thing moves. The argument is as follows. If a partless thing moves, it moves in time; 'but every time is divisible'. There would be, accordingly, in the case of every moving thing whether partitionable or partless some lesser
10   time than that in which the moving thing moves [over] an [interval] equal to itself, and in that [time] it is necessary that it move [over] some lesser [interval], so that there will be something less than the indivisible, and the indivisible will be divided into a lesser [part] and an excess. Thus the indivisible will be divisible. For the moving thing will be divided into some lesser thing which is itself also
15   moving, just as the time too [will be divided] into a lesser time in which it moves [over] the lesser [interval]. Just as earlier he added to the demonstration [the conclusion] that 'only in this way would there be motion of a partless thing, if time were made of nows' [240b31-33], so he says here that 'a partless thing would move only if it were possible for it to move in a now'. For if it were possible to move in a now, a partless thing and none other would be the thing
20   that moves in it. It is obvious from what has been said before that it is not possible to move in a now, but only to have moved.

241a26-b12: 'No change is infinite' to 'so as not to be bounded by limits, is evident.'

[No change is infinite, for a change is from something to something (whether these are contraries or contradictories), and thus between limits. As for local motion, what cannot change cannot be changing: but a thing cannot traverse an infinite distance: hence it cannot be traversing an infinite distance.]

Since it has been proved earlier that having moved precedes every
25   moving and moving [precedes every] having moved, and to those who were listening in an off-hand way this seemed to hypothesize each change as infinite, for this reason he adds this theorem,
1030,1   proving that there is no infinite change. He proves [this] on the basis that every change is 'from something to something', and that these [limits] are bounded. For in fact a [change] in respect to a contradictory pair, even if it is not from a subject [*hupokeimenon*] to a subject,[108] is nevertheless itself too from something to something:

---

[108] In a contradictory pair, the 'subject' or *hupokeimenon* is defined (225a6-7) as 'that which is signified by affirmation', i.e. the positive term (e.g. being), as opposed to the negative term (e.g. not-being).

for it is from not being to being and from being to not being. Not 5
being too is something, if ['something' is] meant in a general way.
And a [change] according to a contradictory pair has as limits and
boundaries [*horos*] the parts of the contradictory pair, [namely]
being and not-being, while a transformation, being a motion, is from
contrary to contrary. Contraries are bounded [or 'defined' (*hôristai*)],
as, for example, white black, hot cold, health sickness, good bad. For
contraries stand furthest apart, and things that stand furthest
apart are bounded [i.e. there is nothing further]. Both increase and 10
diminution also have limits, the former the perfect magnitude [of a
thing] 'according to its proper nature', and the latter the departure
from this [magnitude]. In the case of locomotion the limits are not
everywhere bounded. For in the case of [locomotion] in a straight
[line] up and down are contraries and bounded, but the motion of
animals, [with] which they move in passing about, no longer has
limits that are bounded. Also, revolving locomotion is not from 15
contrary to contrary, for there is no contrary in locomotion in a
circle. For a revolution is from a thing to the same thing, not from
contrary to contrary. For nothing is itself contrary to itself. If,
accordingly, there is not contrariety in every locomotion, [then] by
this [method] it is not possible to prove the boundedness of the
limits: but he proves it by another method that fits in the case of
every change, assuming in advance that what is unable [*adunaton*] 20
to have been cut cannot be being cut. Since 'unable' is said in many
senses (for in fact both what [is done] barely and with difficulty [is
said to be unable to be done], as we have learned in the [logical]
division of the possible and impossible [*adunaton*], and what cannot
[be done] at all [cf. 226b10-12: *de Caelo* 280b12-14]), the latter
meaning of 'unable' is singled out here. And he says that what is 'in
this way unable' to be cut [aorist tense], in the sense that it cannot
be cut [at all], cannot be being cut [present tense]: neither can what 25
is unable to come to be be coming to be. For it would be coming to be
in vain, unless it were able to come to be. But neither god nor nature
does anything in vain [cf. *de Caelo* 271a33]. Perhaps Aristotle says
more exactly that what cannot come to be is not coming to be at
all.[109] For even if those [legendary giants Otus and Ephialtes] put
[Mt.] Ossa on top of [Mt.] Olympus and [Mt.] Pelion on top of [Mt.]
Ossa, this was not the coming-to-be of an ascent to heaven, since it 30
was not possible for that to come to be. For anything that is coming
to be has also the nature to come to be. If these things are true,
[then] what is unable to change cannot be changing into that into

---

[109] Simplicius seems to take *oude holôs to adunaton genesthai gignesthai*, which in
Aristotle means 'nor *in general* (*holôs*) for what is unable to come to be to be coming to
be', as though it meant 'nor for what is unable to come to be to be coming to be *at all*
(*holôs*)', and to interpret accordingly.

which it was unable to change, so that the converse of this is also
1031,1 true, that if something is changing [present tense] into something, it
is possible for this also to change [aorist tense], that is to have
changed [perfect tense], into that. What can change, that is have
changed, into something, would not be moving [over] an infinite
interval or [with] an infinite motion. For in an infinite motion there
is no extreme into which what is changing will be able to change.
5 Consequently, every motion in respect to place too is finite. For in
fact the same thing from which it began is a kind of limit of a
[motion] in a circle. Having proved these things, accordingly, he
concluded the argument by saying that it is evident that there is not
'an infinite change' in this way, 'so as not to be bounded by limits',
since what is changing is unable not to change into that into which it
is changing.

10   **241b12-20**: 'But whether it can in this way' to the end [of the
book].

[No single motion, except motion in a circle, can be infinite in
an infinite time.]

Having proved that no motion is infinite in this way, so as not to
have as limits that from which and that to which, in addition to
these things he inquires whether it is possible for there to arise a
motion infinite in time that is the same and one in type, by occurring
again and again, since the infinite on a straight [line] was proved to
15 be impossible. But it is possible, accordingly, for there to be a
continuous motion, one in number, by the same [motion] occurring
again and again. And he remarked nicely that nothing prevents a
[motion] that is not one nor the same from being infinite in time, if
one [motion] should succeed another, 'for example if there should be
a transformation after locomotion and after' this 'increase and
[then] again coming to be'. But a [motion] made of these does not
20 become one either in number or in type, as he himself will prove. For
a [motion] that occurs in this way too might be said rather generally
to be one, but it is not properly one and continuous. But we are
inquiring whether it is possible for there to arise a single and
continous motion to infinity by the same [motion occurring] again
and again, not interrupted by some halts in between or composed of
[motions] of dissimilar type, but rather remaining the same in type.
25 He says that one such [motion] alone can arise, [namely] that in a
circle, because in other motions the return to the same things again
and again is interrupted by halts. Only [motion] in a circle,
consequently, can occur as one and continuous [motion], infinite in

time. He added 'in time' nicely, because it is not [infinite] in the
magnitude [over which it occurs] or in the extent of the motion
toward one thing. But these things he announced in advance here,
already in labour with the truth in them, but he will demonstrate   30
them later [cf. 261b27-265b16].

Let this, accordingly, be the clarification of the things in this book
part by part, according to my ability, at all events, and let a concise
summary covering the chief [matters] be something like this: having
assumed, on the basis of the things that were distinguished in
advance, what a continuous thing is and that a magnitude is   1032,1
continuous, he proves first in the case of a line that it is not
composed of partless things, that is of points, nor is it divided into
these, and then that both motion and time, which are continuous
things in a similar way to magnitude, similarly have [the property
of] not being composed of partless things or being divided into
partless things. Having first proved the similarity of magnitude and   5
motion, he then adds time. Wishing also to prove the above-
mentioned things through an illustration, he proves in advance that
a faster thing will traverse a greater [interval] than a slower thing
in an equal time, and an equal [interval] or more in a lesser [time].
He proves this through the definition of faster and through an
illustration. Using these things in addition, he infers through an   10
illustration that both the magnitude and the time are divided to
infinity, and that the faster thing divides the time, while the slower
thing [divides] the magnitude. He proves the same things also from
the fact that something of uniform speed will always traverse half
[an interval] in half the time, and, simply, a lesser [interval] in a
lesser [time]. For if the same divisions occur, [then] if one is divisible
to infinity, the other will be such as well. Having thus proved that   15
the division is into forever divisibles, he solves the argument of Zeno
that attempts to deny motion through the division of magnitudes to
infinity: for if, [Zeno] says, there is motion, it is necessary to traverse
an infinite number of things in a finite time. [Aristotle] proves,
accordingly, that [it will not traverse] an infinite number of things in
a finite [time], but rather in a time that possesses infinity in a
similar way. For in fact just as the magnitude, so too the time   20
possesses infinity, not actually,[110] but rather by virtue of its being
by division to infinity. Having proved that the parts of the
magnitude and the time and, obviously, of the motion too are
divisible, he proves next that a now, properly, even if it seems to be a
part of time, is not a part but rather a limit and a beginning, and
that the limit of the past and the beginning of the future are one and
the same in number. He proves it through having proved that these   25

---

[110] Supplying the iota subscript to *energeiâi*, accidentally omitted in Diels' text.

cannot be two either in the sense of touching (for partless things do not touch), or in the sense of being separate (for they would have time in between). As consequent upon the fact that the now is one and the same he adds that it is also indivisible, and using in addition what has been proved he proves that in a now it is not

30   possible for something either to move or to be at rest, but rather both what is moving moves in time and what is at rest is at rest in time. Next he proves again that not only are motion and time divisible, but also the moving magnitude itself, according to another argument as follows: if it is necessary that what is moving[111] from something to something have some of itself in that from which it is

35   moving, and some in that to which it is moving, what is moving would be divisible. He proves that the motion too is divisible not only in respect to the time, as previously, but also in respect to the

1033,1   magnitude which had been proved divisible immediately before. And having proved thus that that in respect to which the motion [occurs], whether in respect to place or quantity or quality, and the motion and the moving and the time are divisible, he proves next that there are the same divisions of all these things and that all are

5   co-divided similarly with one another, and first that the time and motion are, then that the motion and the moving, and then that those things in respect to which the motion [occurs] are too. Having proposed to prove next that what has changed has changed not in time, but rather in a now that is atomic, he first proves that what has changed is in that in which first it has changed, and thus that that in which first something has changed is atomic and not time.

10   He [logically] divides that in which first [something has changed] in two ways, into the beginning of a change and into the limit, and he proves that there is a limit, but there is not a beginning. There is neither a beginning of a change, nor again of what has changed is there some first [bit] that has changed. Of the things in respect to which a change [occurs], in the case of place and quantity it is not

15   possible to take a first [bit], because they are continuous, while in the case of quality in its own right it is not possible, but incidentally it is possible [sc. to divide it].[112] Nor again in the case of the time in which a change [occurs] is it possible to take a first [bit]. As following upon what has been said, he adds that every thing that is moving has moved earlier and [everything] that has moved was moving earlier. And he proves that also in the case of coming to be and perishing the same argument will fit, and he infers next that [it will

20   fit] also in the case of every change. To these things he adds that it is

---

[111] Reading *anankê to ek tinos*, etc., with CM, instead of omitting *to* with the other MSS and Diels.

[112] Simplicius is either careless or too condensed here, since he seems to suggest that it is possible to take a first bit of a quality incidentally; cf. 988,25-6, 989,16-26.

impossible to be moving [over] a finite line or [with a finite] motion
in an infinite time, unless it is the same one again and again as in
the case of revolution, but not to be coming to be or perishing [in an
infinite time]. Consequent upon this is that it is not possible in a
finite time to move or to be brought to rest either [over] an infinite
interval or [with an infinite] motion or rest.[113] And next he proves 25
that neither will an infinite magnitude traverse a finite interval in a
finite time. And having thus systematically discussed the properties
concerning perfect motions that are exactly completed,[114] he offers
next what is coming to a halt and on this account said to be halting,
proving that even if this is said to be halting, it is the same as what
is moving, and that as many things as are properties of what is
moving will be properties also of what is said in this way to be 30
halting, so that [it is the case] both that it halts in time and that it is
not possible to take a first [bit] in halting just as it is not [possible] in
moving. Having proved these things in the case of motion simply [so
called] and that [so] called in respect to halting, he proves also in the
case of rest as opposed to the whole [of] motion [i.e. including
halting] that in this case too it is in time and that it is not possible to
take a first [bit]. Next he proves what is even more amazing than 35
these things, [namely] that it is impossible for a moving thing, in the
first time in which it moves, to be at any [bit] of the interval in which 1034,1
it is moving such that in the time as a whole both it and each of its
parts are in an interval that is the same as and equal to themselves,
but rather in a now only is it in an interval equal to itself. On the
basis of this he also solved Zeno's argument that says that if an
arrow that is moving locally is always at something equal to itself, 5
and what is at something equal to itself for some time is at rest,
[then] the arrow moving locally is at rest while it is moving. He
solves it through the fact that it is not at something equal to itself in
time, but rather in a now, in which it neither moves any nor is at
rest. There being four arguments of Zeno concerning motion, the
first said that if there is motion, it is necessary that a moving thing 10
will entirely traverse an infinite number of things in a finite time,
which [argument] he already solved. Second is the the one called
'Achilles', this one too stating that if there is motion, the slowest will
not be overtaken by the fastest because of the cutting of magnitudes
to infinity. Having started out from the same [premises], it too will
be solved by the same [arguments]. The third argument of Zeno is
that from the arrow moving locally, and the fourth that from equal 15
magnitudes of equal speed in a stadium, some moving past things

[113] Simplicius is inexact in speaking of 'being brought to rest' with or through 'a rest'
(*êremia*); the proper term would be 'a resting' (*êremêsis*), cf. 1002,17-20.
[114] On exactly completed motions, cf. 1001,8-19 (Simplicius' summary of the
argument immediately preceding the discussion of resting or halting).

that are standing still and others past things that are moving
opposite, and in this way seeming to yield the absurdity that of
magnitudes that are equal and of equal speed one moves [with]
double the motion of the other in the same time and that the same
and equal time is simultaneously double and half. The solution of
20 this [argument] is more manifest. Having set out to resolve in
general the arguments that raise objections against motion, after
those of Zeno he brings in others too and resolves them, of which the
first is that from contradictory pairs. For if what is changing is
neither in that from which it is changing (for it would not yet be
changing; nor in that to which it is changing (for it would no longer
25 be changing), if there is change, [then] there will be something
between the contradictory pair of being and not-being, white and not
white, which indeed is impossible. Consequently, there will not be
change. He solves this argument too from the fact that it is in
neither as a whole, but rather some of it is in that from which, and
some in that to which. He resolves a second and different argument
which seems to raise an objection against motion in this way: if the
things that seem to move [with] the first and most important
30 motion, [namely] the circle and the sphere, do not move but are at
1035,1 rest, since they are in the same place, there would not be motion. He
solves it by proving that things that move in this way are neither
themselves nor their parts in the same place for any [stretch of]
time. After these things he proves that a partless thing does not
move except incidentally by being in a moving body or magnitude.
5 He proves it through three arguments. On top of these, he proves
that no change is infinite, even if it is necessary that there always be
having moved before moving and moving before having moved. He
proves it in respect to each type of change, since the passage is from
what is bounded to what is bounded. And at the end he inquires
whether it is possible for there to arise a motion infinite in time that
10 is the same and one in form, by occurring again and again. And he
says that [motion] in a circle only can be such, since what [occurs]
again and again in it [i.e. motion in a circle] is not interrupted by a
halt. But he will demonstrate this later.

Departures from Diels' text

924,14	reading *triôn* instead of *tria*.
928,21	reading *tína* instead of *tina*.
928,30	inserting *ei men oun grammê* before *ouk*.
929,26	reading *ho* instead of *hê*.
933,10	reading *to* instead of *tôi*.
933,14	reading *kineisthai* instead of *kinoumenon*.
933,27-28	closing parenthesis after *ekineito* instead of after *eis ho*.
936,10	closing parenthesis after *kinoumenon* instead of after *kinêsetai*.
938,7	reading *apodeiknusin* instead of *apodeiknunai*.
940,9	inserting *êtoi tôi* before *ZH*.
943,33	reading *tas* instead of *pas*.
944,8	reading *esti* instead of *ésti*.
944,29	deleting *epeidê* (and eliminating lacuna).
949,16-17	reading *AB...BE* instead of *BE...AB*.
949,17	reading *heautou* instead of *autou*.
950,28	reading *C* instead of *CD*.
956,14	inserting diaeresis after *einai* to indicate anacoluthon.
956,15	deleting *ta*.
956,18-19	reading *tauton, hama phaneron* instead of *tauton hama, phaneron*.
957,9	deleting obelus.
959,21	inserting *palin en tôi mellonti* after *autou* (and deleting *<mellonti palin en tôi>*.
959,22	reading *hôste* instead of *kai houtô*.
959,26	deleting *<nun>*.
960,18-19	closing parenthesis after *adiaireton* instead of after *hêmera*.
964,16-19	opening quotation before *kai ek tôn nun* instead of before *touto de*.
964,28	reading *ti dunamei* instead of *to dunamei*.
968,8	reading *kata tauton* instead of *kat'auta*.
971,4	correcting *oti* to *hoti*.
973,1	inserting *sunekhes einai* after *to*.
973,5-6	reading *tou AC kinêseis* instead of *tês AC kinêseôs*.
973,27-28	reading query after *kineisthai* instead of raised stop, and raised stop after *outhen* instead of query.
976,16	reading *hês parousia to C kineisthai* instead of *hês parousiâi tôi C to kineisthai*.
977,12	reading *parousia* instead of *parousiâi*.
977,27-28	closing parenthesis after *estin* instead of after *poiotês*.
980,26	inserting *ouk* after *pou*.
982,14	deleting *<ho>*.
985,25	inserting *ê en amphoterois* after *en oudeterôi*.

986,6   deleting <*ta*>.
989,29   reading *oude en toutois* instead of *en toutois*.
992,4   reading *oute en allôi* instead of *hoti en allôi*.
998,1-2   reading *to gegonenai tou ginesthai oude to ginesthai tou gegonenai* instead of *tou gegonenai to ginesthai oude tou ginesthai to gegonenai*.
998,26   deleting *gar*.
1000,1   reading *to...kinoumenon* instead of *tou...kinoumenou*.
1000,27-28   reading comma after *kinoumenon* instead of full stop, and full stop after *kineisthai* instead of comma.
1006,30   deleting *mê*.
1009,16   reading *to êremoun* instead of *to êremein*.
1015,14   reading *diistatai* instead of *te histatai*.
1017,21   omitting full stop after *einai*.
1020,14-15   reading *touto* (bis) instead of *tote*.
1021,11   reading query after *leukon* instead of comma.
1022,6   correcting *autois* to *hautois*.
1023,31   inserting *ta merê* after *tina*.
1024,6   reading *prôton* after *peripherein*, and omitting *prôtou* before *A*.
1024,8   reading *arxamenên* instead of *arxamenôn*.
1026,2   correcting *epiphaneia* to *epiphaneiâi*.
1032,20   correcting *energeia* to *energeiâi*.
1032,33   reading *anankê to ek tinos* instead of *anankê ek tinos*.

Appendix
The Commentators*

The 15,000 pages of the Ancient Greek Commentaries on Aristotle are the largest corpus of Ancient Greek philosophy that has not been translated into English or other modern European languages. The standard edition (*Commentaria in Aristotelem Graeca*, or *CAG*) was produced by Hermann Diels as general editor under the auspices of the Prussian Academy in Berlin. Arrangements have now been made to translate at least a large proportion of this corpus, along with some other Greek and Latin commentaries not included in the Berlin edition, and some closely related non-commentary works by the commentators.

The works are not just commentaries on Aristotle, although they are invaluable in that capacity too. One of the ways of doing philosophy between A.D. 200 and 600, when the most important items were produced, was by writing commentaries. The works therefore represent the thought of the Peripatetic and Neoplatonist schools, as well as expounding Aristotle. Furthermore, they embed fragments from all periods of Ancient Greek philosophical thought: this is how many of the Presocratic fragments were assembled, for example. Thus they provide a panorama of every period of Ancient Greek philosophy.

The philosophy of the period from A.D. 200 to 600 has not yet been intensively explored by philosophers in English-speaking countries, yet it is full of interest for physics, metaphysics, logic, psychology, ethics and religion. The contrast with the study of the Presocratics is striking. Initially the incomplete Presocratic fragments might well have seemed less promising, but their interest is now widely known, thanks to the philological and philosophical effort that has been concentrated upon them. The incomparably vaster corpus which preserved so many of those fragments offers at least as much interest, but is still relatively little known.

The commentaries represent a missing link in the history of philosophy: the Latin-speaking Middle Ages obtained their

* Reprinted from the Editor's General Introduction to the series in Christian Wildberg, *Philoponus Against Aristotle on the Eternity of the World*, London and Ithaca N.Y., 1987.

knowledge of Aristotle at least partly through the medium of the commentaries. Without an appreciation of this, mediaeval interpretations of Aristotle will not be understood. Again, the ancient commentaries are the unsuspected source of ideas which have been thought, wrongly, to originate in the later mediaeval period. It has been supposed, for example, that Bonaventure in the thirteenth century invented the ingenious arguments based on the concept of infinity which attempt to prove the Christian view that the universe had a beginning. In fact, Bonaventure is merely repeating arguments devised by the commentator Philoponus 700 years earlier and preserved in the meantime by the Arabs. Bonaventure even uses Philoponus' original examples. Again, the introduction of impetus theory into dynamics, which has been called a scientific revolution, has been held to be an independent invention of the Latin West, even if it was earlier discovered by the Arabs or their predecessors. But recent work has traced a plausible route by which it could have passed from Philoponus, via the Arabs, to the West.

The new availability of the commentaries in the sixteenth century, thanks to printing and to fresh Latin translations, helped to fuel the Renaissance break from Aristotelian science. For the commentators record not only Aristotle's theories, but also rival ones, while Philoponus as a Christian devises rival theories of his own and accordingly is mentioned in Galileo's early works more frequently than Plato.[1]

It is not only for their philosophy that the works are of interest. Historians will find information about the history of schools, their methods of teaching and writing and the practices of an oral tradition.[2] Linguists will find the indexes and translations an aid for studying the development of word meanings, almost wholly

---

[1] See Fritz Zimmermann, 'Philoponus' impetus theory in the Arabic tradition'; Charles Schmitt, 'Philoponus' commentary on Aristotle's *Physics* in the sixteenth century', and Richard Sorabji, 'John Philoponus', in Richard Sorabji (ed.), *Philoponus and the Rejection of Aristotelian Science* (London and Ithaca, N.Y. 1987).

[2] See e.g. Karl Praechter, 'Die griechischen Aristoteleskommentare', *Byzantinische Zeitschrift* 18 (1909), 516-38; M. Plezia, *de Commentariis Isagogicis* (Cracow 1947); M. Richard, *'Apo Phônês'*, *Byzantion* 20 (1950), 191-222; É. Evrard, *L'Ecole d'Olympiodore et la composition du commentaire à la physique de Jean Philopon*, Diss. (Liège 1957); L.G. Westerink, *Anonymous Prolegomena to Platonic Philosophy* (Amsterdam 1962) (new revised edition, translated into French, Collection Budé, forthcoming); A.-J. Festugière, 'Modes de composition des commentaires de Proclus', *Museum Helveticum* 20 (1963), 77-100, repr. in his *Études* (1971), 551-74; P. Hadot, 'Les divisions des parties de la philosophie dans l'antiquité, *Museum Helveticum* 36 (1979), 201-23; I. Hadot, 'La division néoplatonicienne des écrits d'Aristote', in J. Wiesner (ed.), *Aristoteles Werk und Wirkung* (Paul Moraux gewidmet), vol. 2 (Berlin 1986); I. Hadot, 'Les introductions aux commentaires exégétiques chez les auteurs néoplatoniciens et les auteurs chrétiens', in M. Tardieu (ed.), *Les règles de l'interprétation* (Paris 1987), 99-119. These topics will be treated, and a bibliography supplied, in a collection of articles on the commentators in general.

uncharted in Liddell and Scott's *Lexicon*, and for checking shifts in grammatical usage.

Given the wide range of interests to which the volumes will appeal, the aim is to produce readable translations, and to avoid so far as possible presupposing any knowledge of Greek. Footnotes will explain points of meaning, give cross-references to other works, and suggest alternative interpretations of the text where the translator does not have a clear preference. The introduction to each volume will include an explanation why the work was chosen for translation: none will be chosen simply because it is there. Two of the Greek texts are currently being re-edited – those of Simplicius *in Physica* and *in de Caelo* – and new readings will be exploited by translators as they become available. Each volume will also contain a list of proposed emendations to the standard text. Indexes will be of more uniform extent as between volumes than is the case with the Berlin edition, and there will be three of them: an English-Greek glossary, a Greek-English index, and a subject index.

The commentaries fall into three main groups. The first group is by authors in the Aristotelian tradition up to the fourth century A.D. This includes the earliest extant commentary, that by Aspasius in the first half of the second century A.D. on the *Nicomachean Ethics*. The anonymous commentary on Books 2, 3, 4 and 5 of the *Nichomachean Ethics*, in *CAG* vol. 20, may be partly or wholly by Adrastus, a generation later.[3] The commentaries by Alexander of Aphrodisias (appointed to his chair between A.D. 198 and 209) represent the fullest flowering of the Aristotelian tradition. To his successors Alexander was The Commentator *par excellence*. To give but one example (not from a commentary) of his skill at defending and elaborating Aristotle's views, one might refer to his defence of Aristotle's claim that space is finite against the objection that an edge of space is conceptually problematic.[4] Themistius (*fl.* late 340s to 384 or 385) saw himself as the inventor of paraphrase, wrongly thinking that the job of commentary was completed.[5] In fact, the Neoplatonists were to introduce new dimensions into commentary. Themistius' own relation to the Neoplatonist as opposed to the Aristotelian tradition is a matter of controversy,[6] but it would be

[3] Anthony Kenny, *The Aristotelian Ethics* (Oxford 1978), 37, n.3; Paul Moraux, *Der Aristotelismus bei den Griechen*, vol. 2 (Berlin 1984), 323-30.

[4] Alexander, *Quaestiones* 3.12, discussed in my *Matter, Space and Motion* (London and Ithaca, N.Y. 1988). For Alexander see R.W. Sharples, 'Alexander of Aphrodisias: scholasticism and innovation', in W. Haase (ed.), *Aufstieg und Niedergang der römischen Welt*, part 2 *Principat*, vol. 36.2, *Philosophie und Wissenschaften* (1987).

[5] Themistius *in An. Post.* 1,2-12. See H.J. Blumenthal, 'Photius on Themistius (Cod.74): did Themistius write commentaries on Aristotle?', *Hermes* 107 (1979), 168-82.

[6] For different views, see H.J. Blumenthal, 'Themistius, the last Peripatetic commentator on Aristotle?', in Glen W. Bowersock, Walter Burkert, Michael C.J. Putnam, *Arktouros*, Hellenic Studies Presented to Bernard M.W. Knox, (Berlin and

agreed that his commentaries show far less bias than the full-blown Neoplatonist ones. They are also far more informative than the designation 'paraphrase' might suggest, and it has been estimated that Philoponus' *Physics* commentary draws silently on Themistius six hundred times.[7] The pseudo-Alexandrian commentary on *Metaphysics* 6–14, of unknown authorship, has been placed by some in the same group of commentaries as being earlier than the fifth century.[8]

By far the largest group of extant commentaries is that of the Neoplatonists up to the sixth century A.D. Nearly all the major Neoplatonists, apart from Plotinus (the founder of Neoplatonism), wrote commentaries on Aristotle, although those of Iamblichus (*c.* 250 – *c.* 325) survive only in fragments, and those of three Athenians, Plutarchus (died 432), his pupil Proclus (410–485) and the Athenian Damascius (*c.* 462 – after 538), are lost.[9] As a result of these losses, most of the extant Neoplatonist commentaries come from the late fifth and the sixth centuries and a good proportion from Alexandria. There are commentaries by Plotinus' disciple and editor Porphyry (232 – 309), by Iamblichus' pupil Dexippus (*c.* 330), by Proclus' teacher Syrianus (died *c.* 437), by Proclus' pupil Ammonius (435/445 – 517/526), by Ammonius' three pupils Philoponus (*c.* 490 to 570s), Simplicius (wrote after 532, probably after 538) and Asclepius (sixth century), by Ammonius' next but one successor Olympiodorus (495/505 – after 565), by Elias (*fl.* 541?), by David (second half of the sixth century, or beginning of the seventh) and by Stephanus (took the chair in Constantinople *c.* 610). Further,

---

N.Y., 1979), 391-400; E.P. Mahoney, 'Themistius and the agent intellect in James of Viterbo and other thirteenth-century philosophers: (Saint Thomaas Aquinas, Siger of Brabant and Henry Bate)', *Augustiniana* 23 (1973), 422-67, at 428-31; id., 'Neoplatonism, the Greek commentators and Renaissance Aristotelianism', in D.J. O'Meara (ed.), *Neoplatonism and Christian Thought* (Albany N.Y. 1982), 169-77 and 264-82, esp. n. 1, 264-6; Robert Todd, introduction to translation of Themistius *in DA 3,4-8,* forthcoming in a collection of translations by Frederick Schroeder and Robert Todd of material in the commentators relating to the intellect.

[7] H. Vitelli, *CAG* 17, p. 992, s.v. Themistius.

[8] The similarities to Syrianus (died *c.*437) have suggested to some that it predates Syrianus (most recently Leonardo Tarán, review of Paul Moraux, *Der Aristotelismus*, vol. 1, in *Gnomon* 46 (1981), 721-50 at 750), to others that it draws on him (most recently P. Thillet, in the Budé edition of Alexander *de Fato*, p. lvii). Praechter ascribed it to Michael of Ephesus (eleventh of twelfth century), in his review of *CAG* 22.2, in *Göttingische Gelehrte Anzeiger* 168 (1906), 861-907.

[9] The Iamblichus fragments are collected in Greek by Bent Dalsgaard Larsen, *Jamblique de Chalcis, Exégète et Philosophe* (Aarhus 1972), vol.2. Most are taken from Simplicius, and will accordingly be translated in due course. The evidence on Damascius' commentaries is given in L.G. Westerink, *The Greek Commentaries on Plato's Phaedo*, vol.2., Damascius (Amsterdam 1977), 11-12; on Proclus' in L.G. Westerink, *Anonymous Prolegomena to Platonic Philosophy* (Amsterdam 1962), xii, n.22; on Plutarchus' in H.M. Blumenthal, 'Neoplatonic elements in the de Anima commentaries', *Phronesis* 21 (1976), 75.

a commentary on the *Nicomachean Ethics* has been ascribed speculatively to Ammonius' brother Heliodorus, and there is a commentary by Simplicius' colleague Priscian of Lydia on Aristotle's successor Theophrastus. Of these commentators some of the last were Christians (Philoponus, Elias, David and Stephanus), but they were Christians writing in the Neoplatonist tradition, as was also Boethius who produced a number of commentaries in Latin before his death in 525 or 526.

The third group comes from a much later period in Byzantium. The Berlin edition includes only three out of more than a dozen commentators described in Hunger's *Byzantinisches Handbuch*.[10] The two most important are Eustratius (1050/1060 – c. 1120), and Michael of Ephesus. It has been suggested that these two belong to a circle organised by the princess Anna Comnena in the twelfth century, and accordingly the completion of Michael's commentaries has been redated from 1040 to 1138.[11] His commentaries include areas where gaps had been left. Not all of these gap-fillers are extant, but we have commentaries on the neglected biological works, on the *Sophistici Elenchi*, and a small fragment of one on the *Politics*. The lost *Rhetoric* commentary had a few antecedents, but the *Rhetoric* too had been comparatively neglected. Another product of this period may have been the composite commentary on the *Nicomachean Ethics* (*CAG* 20) by various hands, including Eustratius and Michael, along with some earlier commentators, and an improvisation for Book 7. Whereas Michael follows Alexander and the conventional Aristotelian tradition, Eustratius' commentary introduces Platonist, Christian and anti-Islamic elements.[12]

The composite commentary was to be translated into Latin in the next century by Robert Grosseteste in England. But Latin translations of various logical commentaries were made from the Greek still earlier by James of Venice (*fl. c.* 1130), a contemporary of Michael of Ephesus, who may have known him in Constantinople.

---

[10] Herbert Hunger, *Die hochsprachliche profane Literatur der Byzantiner*, vol.1 (= *Byzantinisches Handbuch*, part 5, vol.1) (Munich 1978), 25-41. See also B.N. Tatakis, *La Philosophie Byzantine* (Paris 1949).

[11] R. Browning, 'An unpublished funeral oration on Anna Comnena', *Proceedings of the Cambridge Philological Society* n.s. 8 (1962), 1-12, esp. 6-7.

[12] R. Browning, op.cit. H.D.P. Mercken, *The Greek Commentaries of the Nicomachean Ethics of Aristotle in the Latin Translation of Grosseteste, Corpus Latinum Commentariorum in Aristotelem Graecorum* VI 1 (Leiden 1973), ch.1, 'The compilation of Greek commentaries on Aristotle's Nicomachean Ethics'. Sten Ebbesen, 'Anonymi Aurelianensis I Commentarium in *Sophisticos Elenchos*', *Cahiers de l'Institut Moyen Age Grecque et Latin* 34 (1979), 'Boethius, Jacobus Veneticus, Michael Ephesius and "Alexander" ', pp. v-xiii; id., *Commentators and Commentaries on Aristotle's Sophistici Elenchi*, 3 parts, *Corpus Latinum Commentariorum in Aristotelem Graecorum*, vol. 7 (Leiden 1981); A. Preus, *Aristotle and Michael of Ephesus on the Movement and Progression of Animals* (Hildesheim 1981), introduction.

And later in that century other commentaries and works by commentators were being translated from Arabic versions by Gerard of Cremona (died 1187).[13] So the twelfth century resumed the transmission which had been interrupted at Boethius' death in the sixth century.

The Neoplatonist commentaries of the main group were initiated by Porphyry. His master Plotinus had discussed Aristotle, but in a very independent way, devoting three whole treatises (*Enneads* 6.1–3) to attacking Aristotle's classification of the things in the universe into categories. These categories took no account of Plato's world of Ideas, were inferior to Plato's classifications in the *Sophist* and could anyhow be collapsed, some of them into others. Porphyry replied that Aristotle's categories could apply perfectly well to the world of intelligibles and he took them as in general defensible.[14] He wrote two commentaries on the *Categories*, one lost, and an introduction to it, the *Isagôgê*, as well as commentaries, now lost, on a number of other Aristotelian works. This proved decisive in making Aristotle a necessary subject for Neoplatonist lectures and commentary. Proclus, who was an exceptionally quick student, is said to have taken two years over his Aristotle studies, which were called the Lesser Mysteries, and which preceded the Greater Mysteries of Plato.[15] By the time of Ammonius, the commentaries reflect a teaching curriculum which begins with Porphyry's *Isagôgê* and Aristotle's *Categories*, and is explicitly said to have as its final goal a (mystical) ascent to the supreme Neoplatonist deity, the One.[16] The curriculum would have progressed from Aristotle to Plato, and would have culminated in Plato's *Timaeus* and *Parmenides*. The latter was read as being about the One, and both works were established in this place in the curriculum at least by

---

[13] For Grosseteste, see Mercken as in n. 12. For James of Venice, see Ebbesen as in n. 12, and L. Minio-Paluello, 'Jacobus Veneticus Grecus', *Traditio* 8 (1952), 265-304; id., 'Giacomo Veneto e l'Aristotelismo Latino', in Pertusi (ed.), *Venezia e l'Oriente fra tardo Medioevo e Rinascimento* (Florence 1966), 53-74, both reprinted in his *Opuscula* (1972). For Gerard of Cremona, see M. Steinschneider, *Die europäischen Übersetzungen aus dem arabischen bis Mitte des 17. Jahrhunderts* (repr. Graz 1956); E. Gilson, *History of Christian Philosophy in the Middle Ages* (London 1955), 235-6 and more generally 181-246. For the translators in general, see Bernard G. Dod, 'Aristoteles Latinus', in N. Kretzmann, A. Kenny, J. Pinborg (eds). *The Cambridge History of Latin Medieval Philosophy* (Cambridge 1982).

[14] See P. Hadot, 'L'harmonie des philosophies de Plotin et d'Aristote selon Porphyre dans le commentaire de Dexippe sur les Catégories', in *Plotino e il neoplatonismo in Oriente e in Occidente* (Rome 1974), 31-47; A.C. Lloyd, 'Neoplatonic logic and Aristotelian logic', *Phronesis* 1 (1955-6), 58-79 and 146-60.

[15] Marinus, *Life of Proclus* ch.13, 157,41 (Boissonade).

[16] The introductions to the *Isagôgê* by Ammonius, Elias and David, and to the *Categories* by Ammonius, Simplicius, Philoponus, Olympiodorus and Elias are discussed by L.G. Westerink, *Anonymous Prolegomena* and I. Hadot, 'Les Introductions', see n. 2. above.

the time of Iamblichus, if not earlier.[17]

Before Porphyry, it had been undecided how far a Platonist should accept Aristotle's scheme of categories. But now the proposition began to gain force that there was a harmony between Plato and Aristotle on most things.[18] Not for the only time in the history of philosophy, a perfectly crazy proposition proved philosophically fruitful. The views of Plato and of Aristotle had both to be transmuted into a new Neoplatonist philosophy in order to exhibit the supposed harmony. Iamblichus denied that Aristotle contradicted Plato on the theory of Ideas.[19] This was too much for Syrianus and his pupil Proclus. While accepting harmony in many areas,[20] they could see that there was disagreement on this issue and also on the issue of whether God was causally responsible for the existence of the ordered physical cosmos, which Aristotle denied. But even on these issues, Proclus' pupil Ammonius was to claim harmony, and, though the debate was not clear cut,[21] his claim was on the whole to prevail. Aristotle, he maintained, accepted Plato's Ideas,[22] at least in the form of principles (*logoi*) in the divine intellect, and these principles were in turn causally responsible for the beginningless existence of the physical universe. Ammonius wrote a whole book to show that Aristotle's God was thus an efficient cause, and though the book is lost, some of its principal arguments are preserved by Simplicius.[23] This tradition helped to make it possible for Aquinas to claim Aristotle's God as a Creator, albeit not in the sense of giving

---

[17] Proclus *in Alcibiadem 1* p.11 (Creuzer); Westerink, *Anonymous Prolegomena*, ch. 26, 12f. For the Neoplatonist curriculum see Westerink, Festugière, P. Hadot and I. Hadot in n. 2.

[18] See e.g. P. Hadot (1974), as in n. 14 above; H.J. Blumenthal, 'Neoplatonic elements in the de Anima commentaries', *Phronesis* 21 (1976), 64-87; H.A. Davidson, 'The principle that a finite body can contain only finite power', in S. Stein and R. Loewe (eds), *Studies in Jewish Religious and Intellectual History presented to A. Altmann* (Alabama 1979), 75-92; Carlos Steel, 'Proclus et Aristote', Proceedings of the Congrès Proclus held in Paris 1985, J. Pépin and H.D. Saffrey (eds), *Proclus, lecteur et interprète des anciens* (Paris 1987), 213-25; Koenraad Verrycken, *God en Wereld in de Wijsbegeerte van Ioannes Philoponus*, Ph.D. Diss. (Louvain 1985).

[19] Iamblichus ap. Elian *in Cat.* 123,1-3.

[20] Syrianus *in Metaph.* 80,4-7; Proclus *in Tim.* 1.6,21-7,16.

[21] Asclepius sometimes accepts Syranius' interpretation (*in Metaph*, 433,9-436,6); which is, however, qualified, since Syrianus thinks Aristotle is really committed willy-nilly to much of Plato's view (*in Metaph*, 117,25-118,11; ap. Asclepium *in Metaph.* 433,16; 450,22); Philoponus repents of his early claim that Plato is not the target of Aristotle's attack, and accepts that Plato is rightly attacked for treating ideas as independent entities outside the divine Intellect (*in DA* 37,18-31; *in Phys.* 225,4-226,11; *contra Procl.* 26,24-32,13; *in An. Post.* 242,14–243,25).

[22] Asclepius *in Metaph* from the voice of (i.e. from the lectures of) Ammonius 69,17-21; 71,28; cf. Zacharias *Ammonius, Patrologia Graeca* vol. 85, col. 952 (Colonna).

[23] Simplicius *in Phys.* 1361,11-1363,12. See H.A. Davidson; Carlos Steel; Koenraad Verrycken in n.18 above.

the universe a beginning, but in the sense of being causally responsible for its beginningless existence.[24] Thus what started as a desire to harmonise Aristotle with Plato finished by making Aristotle safe for Christianity. In Simplicius, who goes further than anyone,[25] it is a formally stated duty of the commentator to display the harmony of Plato and Aristotle in most things.[26] Philoponus, who with his independent mind had thought better of his earlier belief in harmony, is castigated by Simplicius for neglecting this duty.[27]

The idea of harmony was extended beyond Plato and Aristotle to Plato and the Presocratics. Plato's pupils Speusippus and Xenocrates saw Plato as being in the Pythagorean tradition.[28] From the third to first centuries B.C., pseudo-Pythagorean writings present Platonic and Aristotelian doctrines as if they were the ideas of Pythagoras and his pupils,[29] and these forgeries were later taken by the Neoplatonists as genuine. Plotinus saw the Presocratics as precursors of his own views,[30] but Iamblichus went far beyond him by writing ten volumes on Pythagorean philosophy.[31] Thereafter Proclus sought to unify the whole of Greek philosophy by presenting it as a continuous clarification of divine revelation,[32] and Simplicius argued for the same general unity in order to rebut Christian charges of contradictions in pagan philosophy.[33]

Later Neoplatonist commentaries tend to reflect their origin in a teaching curriculum:[34] from the time of Philoponus, the discussion is often divided up into lectures, which are subdivided into studies of doctrine and of text. A general account of Aristotle's philosophy is prefixed to the *Categories* commentaries and divided, according to a formula of Proclus,[35] into ten questions. It is here that commentators explain the eventual purpose of studying Aristotle (ascent to the One) and state (if they do) the requirement of

[24] See Richard Sorabji, *Matter, Space and Motion* (London and Ithaca N.Y. 1988), ch. 15.
[25] See e.g. H.J. Blumenthal in n. 18 above.
[26] Simplicius *in Cat.* 7,23-32.
[27] Simplicius *in Cael.* 84,11-14; 159,2-9. On Philoponus' *volte face* see n. 21 above.
[28] See e.g. Walter Burkert, *Weisheit und Wissenschaft* (Nürnberg 1962), translated as *Lore and Science in Ancient Pythagoreanism* (Cambridge Mass. 1972), 83-96.
[29] See Holger Thesleff, *An Introduction to the Pythagorean writings of the Hellenistic Period* (Åbo 1961); Thomas Alexander Szlezák, *Pseudo-Archytas über die Kategorien*, Peripatoi vol. 4 (Berlin and New York 1972).
[30] Plotinus e.g. 4.8.1; 5.1.8 (10-27); 5.1.9.
[31] See Dominic O'Meara, *Pythagoras Revived: Mathematics and Philosophy in late Antiquity*, forthcoming.
[32] See Christian Guérard, 'Parménide d'Elée selon les Néoplatoniciens', forthcoming.
[33] Simplicius *in Phys.* 28,32-29,5; 640,12-18. Such thinkers as Epicurus and the Sceptics, however, were not subject to harmonisation.
[34] See the literature in n. 2 above.   [35] ap. Elian *in Cat.* 107,24-6.

displaying the harmony of Plato and Aristotle. After the ten-point introduction to Aristotle, the *Categories* is given a six-point introduction, whose antecedents go back earlier than Neoplatonism, and which requires the commentator to find a unitary theme or scope (*skopos*) for the treatise. The arrangements for late commentaries on Plato are similar. Since the Plato commentaries form part of a single curriculum they should be studied alongside those on Aristotle. Here the situation is easier, not only because the extant corpus is very much smaller, but also because it has been comparatively well served by French and English translators.[36]

Given the theological motive of the curriculum and the pressure to harmonise Plato with Aristotle, it can be seen how these commentaries are a major source for Neoplatonist ideas. This in turn means that it is not safe to extract from them the fragments of the Presocratics, or of other authors, without making allowance for the Neoplatonist background against which the fragments were originally selected for discussion. For different reasons, analogous warnings apply to fragments preserved by the pre-Neoplatonist commentator Alexander.[37] It will be another advantage of the present translations that they will make it easier to check the distorting effect of a commentator's background.

Although the Neoplatonist commentators conflate the views of Aristotle with those of Neoplatonism, Philoponus alludes to a certain convention when he quotes Plutarchus expressing disapproval of Alexander for expounding his own philosophical doctrines in a commentary on Aristotle.[38] But this does not stop Philoponus from later inserting into his own commentaries on the *Physics* and *Meteorology* his arguments in favour of the Christian view of Creation. Of course, the commentators also wrote independent works of their own, in which their views are expressed independently of the exegesis of Aristotle. Some of these independent works will be included in the present series of translations.

The distorting Neoplatonist context does not prevent the commentaries from being incomparable guides to Aristotle. The

---

[36] English: Calcidius *in Tim.* (parts by van Winder; den Boeft); Iamblichus fragments (Dillon); Proclus *in Tim.* (Thomas Taylor); Proclus *in Parm.* (Dillon); Proclus *in Parm.*, end of 7th book, from the Latin (Klibansky, Labowsky, Anscombe); Proclus *in Alcib. 1* (O'Neill); Olympiodorus and Damascius *in Phaedonem* (Westerink); Damascius *in Philebum* (Westerink); *Anonymous Prolegomena to Platonic Philosophy* (Westerink). See also extracts in Thomas Taylor, *The Works of Plato*, 5 vols. (1804). French: Proclus *in Tim.* and *in Rempublicam* (Festugière); *in Parm.* (Chaignet); Anon. *in Parm.* (P. Hadot); Damascius *in Parm.* (Chaignet).

[37] For Alexander's treatment of the Stoics, see Robert B. Todd, *Alexander of Aphrodisias on Stoic Physics* (Leiden 1976), 24-9.

[38] Philoponus *in DA* 21,20-3.

introductions to Aristotle's philosophy insist that commentators must have a minutely detailed knowledge of the entire Aristotelian corpus, and this they certainly have. Commentators are also enjoined neither to accept nor reject what Aristotle says too readily, but to consider it in depth and without partiality. The commentaries draw one's attention to hundreds of phrases, sentences and ideas in Aristotle, which one could easily have passed over, however often one read him. The scholar who makes the right allowance for the distorting context will learn far more about Aristotle than he would be likely to on his own.

The relations of Neoplatonist commentators to the Christians were subtle. Porphyry wrote a treatise explicitly against the Christians in 15 books, but an order to burn it was issued in 448, and later Neoplatonists were more circumspect. Among the last commentators in the main group, we have noted several Christians. Of these the most important were Boethius and Philoponus. It was Boethius' programme to transmit Greek learning to Latin-speakers. By the time of his premature death by execution, he had provided Latin translations of Aristotle's logical works, together with commentaries in Latin but in the Neoplatonist style on Porphyry's *Isagôgê* and on Aristotle's *Categories* and *de Interpretatione*, and interpretations of the *Prior* and *Posterior Analytics, Topics* and *Sophistici Elenchi*. The interruption of his work meant that knowledge of Aristotle among Latin-speakers was confined for many centuries to the logical works. Philoponus is important both for his proofs of the Creation and for his progressive replacement of Aristotelian science with rival theories, which were taken up at first by the Arabs and came fully into their own in the West only in the sixteenth century.

Recent work has rejected the idea that in Alexandria the Neoplatonists compromised with Christian monotheism by collapsing the distinction between their two highest deities, the One and the Intellect. Simplicius (who left Alexandria for Athens) and the Alexandrians Ammonius and Asclepius appear to have acknowledged their beliefs quite openly, as later did the Alexandrian Olympiodorus, despite the presence of Christian students in their classes.[39]

The teaching of Simplicius in Athens and that of the whole pagan Neoplatonist school there was stopped by the Christian Emperor Justinian in 529. This was the very year in which the Christian

[39] For Simplicius, see I. Hadot, *Le Problème du Néoplatonisme Alexandrin: Hiéroclès et Simplicius* (Paris 1978); for Ammonius and Asclepius, Koenraad Verrycken, *God en Wereld in de Wijsbegeerte van Ioannes Philoponus*, Ph.D. Diss. (Louvain 1985); for Olympiodorus, L.G. Westerink, *Anonymous Prolegomena to Platonic Philosophy* (Amsterdam 1962).

Philoponus in Alexandria issued his proofs of Creation against the earlier Athenian Neoplatonist Proclus. Archaeological evidence has been offered that, after their temporary stay in Ctesiphon (in present-day Iraq), the Athenian Neoplatonists did not return to their house in Athens, and further evidence has been offered that Simplicius went to Ḥarrān (Carrhae), in present-day Turkey near the Iraq border.[40] Wherever he went, his commentaries are a treasure house of information about the preceding thousand years of Greek philosophy, information which he painstakingly recorded after the closure in Athens, and which would otherwise have been lost. He had every reason to feel bitter about Christianity, and in fact he sees it and Philoponus, its representative, as irreverent. They deny the divinity of the heavens and prefer the physical relics of dead martyrs.[41] His own commentaries by contrast culminate in devout prayers.

Two collections of articles by various hands are planned, to make the work of the commentators better known. The first is devoted to Philoponus;[42] the second will be about the commentators in general, and will go into greater detail on some of the issues briefly mentioned here.[43]

[40] Alison Frantz, 'Pagan philosophers in Christian Athens', *Proceedings of the American Philosophical Society* 119 (1975), 29-38; M. Tardieu, 'Témoins orientaux du *Premier Alcibiade* à Ḥarrān et à Nag 'Hammādi', *Journal Asiatique* 274 (1986); id., 'Les calendriers en usage à Ḥarrān d'après les sources arabes et le commentaire de Simplicius à la *Physique* d'Aristote', in I. Hadot (ed.), *Simplicius, sa vie, son oeuvre, sa survie* (Berlin 1987), 40-57; *Coutumes nautiques mésopotamiennes chez Simplicius*, in preparation. The opposing view that Simplicius returned to Athens is most fully argued by Alan Cameron, 'The last days of the Academy at Athens', *Proceedings of the Cambridge Philological Society* 195, n.s. 15 (1969), 7-29.

[41] Simplicius *in Cael.* 26,4-7; 70,16-18; 90,1-18; 370,29-371,4. See on his whole attitude Philippe Hoffmann, 'Simplicius' polemics', in Richard Sorabji (ed.), *Philoponus and the Rejection of Aristotelian Science* (London and Ithaca, N.Y. 1987).

[42] Richard Sorabji (ed.), *Philoponus and the Rejection of Aristotelian Science* (London and Ithaca, N.Y. 1987).

[43] The lists of texts and previous translations of the commentaries included in Wildberg, *Philoponus Against Aristotole on the Eternity of the World* (pp.12ff.) are not included here. The list of translations should be augmented by: F.L.S. Bridgman, Heliodorus (?) in *Ethica Nicomachea*, London 1807.

I am grateful for comments to Henry Blumenthal, Victor Caston, I. Hadot, Paul Mercken, Alain Segonds, Robert Sharples, Robert Todd, L.G. Westerink and Christian Wildberg.

Bibliography

Becker, O., 'Formallogisches und Mathematisches in griechischen philosophischen Texten', *Philologus* 100, 1956, 108-12.

————, 'Zum Text eines mathematischen Beweises im eudemischen Bericht über die Quadraturen der Möndchen durch Hippocrates von Chios bei Simplikios (Arist. *Phys.* S. 66 Diels)', *Philologus* 99, 1955, 313-16.

Bormann, K., 'The interpretation of Parmenides by the Neoplatonist Simplicius', *Monist* 62, 1979, 30-42.

Borovskij, J., 'Specilegium Simplicianum (in Arist. *Phys.* II, quattuor priores, ed. Diels, p. 679, 12 sq.)', *Studi Classici in onore di Quinto Cataudella*, Catania, Facc. di Lett. e Filos., 1972, vol. 2, 209-10.

Buhle, J.G., 'De Simplicio vita, ingenio et meritis', *Göttingische Gelehrte Anzeigen*, 1786, 1977f.

Cameron, A., 'The last days of the Academy at Athens', *Proceedings of the Cambridge Philological Society* 15, 1969, 7-29.

Carteron, H., 'Does Aristotle have a mechanics?', in J. Barnes, R. Sorabji and M. Schofield (eds), *Articles on Aristotle*, vol. 1: *Science*, London 1975, 161-74.

Caveing, M., *Zénon d'Élée. Prolégomènes aux doctrines du continu. Études historiques et critiques des Fragments et Témoignages (= Histoire des doctrines de l'antiquité classique* 7), Paris 1982.

Clagett, M., 'A thirteenth-century fragment of Simplicius' commentary on the *Physics* of Aristotle, Quadratura per lunulas', in J.H. Mundy, R.W. Emery and B.N. Nelson (eds), *Essays in Medieval Life and Thought presented in honour of A.P. Evans*, New York 1955, 99-108.

Cordero, N.L., 'Analyse de l'édition Aldine du commentaire de Simplicius à la *Physique* d'Aristote', *Hermes* 105, 1977, 42-54.

————, 'Les sources vénitiennes de l'edition Aldine Livre I du Commentaire de Simplicius sur "la Physique" d'Aristote', *Scriptorium* 39 (1) 1985, 70-88.

Coxon, A.H., 'The manuscript tradition of Simplicius' commentary on Aristotle's *Physics* i-iv', *Classical Quarterly* 18, 1968, 70-5.

Dodds, E.R., 'Simplicius', *Oxford Classical Dictionary*, Oxford 1970.

Ducci, E., 'Il *to eon* Parmenideo nella interpretazione di Simplicio', *Angelicum* 40, 1963, 173-94, 313-27.

Evangelides, M., 'Ho Zenonos peri tou apeirou to megethos logos (Simpl. *in Phys.* 141.1 Diels)', *Philosophika Meletêmata*, Athens 1885, 78-96.

Frantz, A., 'Pagan philosophers in Christian Athens', *Proceedings of the American Philosophical Society* 119, 1975, 29-38.

Gätje, H., 'Simplikios in der arabischen Überlieferung', *Der Islam* 59, 1982, 6-31.

Genequand, C., 'Quelques aspects de l'idée de nature d'Aristote à Al-Ghazali', *Revue de Theologie et de Philosophie* 116, 1984, 105-29.

Gigante, M., 'Sul testo del Misoumenos di Menandro', *Bolletino del Comitato per la Preparazione dell'Edizione nazionale dei Classici greci e latini* 14, 1966, 13-21.

Gomperz, T., 'Anaxagoras (bei Simpl.. Phys. 139,11 Diels)', *Beiträge zur Kritik und Erklärung griechischer Schriftsteller* IV (=*Sitzungsberichte der philo.-hist. Classe d. k. Akad. d.Wissenschaften in Wien* 122), Wien 1890.

Hadot, I., 'Les *Introductions* aux commentaires exégétiques chez les auteurs néoplatoniciens et les auteurs chrétiens', in M. Tardieu (ed.), *Les règles de l'interprétation*, Paris 1987.

————, *Simplicius, sa vie, son oeuvre, sa survie*, Berlin 1987.

————, 'La vie et l'oeuvre de Simplicius d'après des sources grecques et arabes', in id. (ed.), *Simplicius, sa vie, son oeuvre, sa survie*, Berlin 1987, 3-29.

————, *Le problème du néoplatonism alexandrin: Hiéroclès et Simplicius*, Paris 1978.

Hoffman, P., 'Simplicius' polemics', in R. Sorabji (ed.), *Philoponus and the Rejection of Aristotelian Science*, London and Ithaca, N.Y. 1987, 57-83.

————, 'Paratasis. De la description aspectuelle des verbes grecs à une définition du temps dans le néoplatonism tardif', *Revue des Études Grecques* 96, 1983, 1-26.

————, 'Jamblique exégète du pythagoricien Archytas: trois originalités d'une doctrine du temps', *Études Philosophiques*, 1980, 307-23.

————, 'Les catégories "où" et "quand" chez Aristote et Simplicius', in P. Aubenque (ed.), *Concepts et catégories dans la pensée antique*, Paris 1980, 217-45.

————, 'Simplicius: Corollarium de Loco', *L'Astronomie dans l'Antiquité Classique*, Actes du colloque tenu à l'Université de Toulouse-Le Mirail, 21-33 Oct. 1977, Paris, Les Belles Letters, 1979, 143-63.

King, H.R., 'Aristotle's Theory of ΤΟΠΟΣ', *Classical Quarterly* 44, 1950, 76-96.

Lloyd, A.C., 'Simplicius' in P. Edwards (ed.), *Encyclopaedia of Philosophy*, vol. 7, 448-9.

Lloyd, G.E.R., 'Saving the appearances', *Classical Quarterly* 28, 1978, 202-22.

Magris, A., 'Archita e l'eterno ritorno', *Elenchos* 3, 1982, 237-58.

Mahoney, E.P., 'Marcilio Ficino's influence on Nicoleto Vernia, Augustino Nifo and Marcantonio Zimara', in G.C. Giamparini (ed.), *Marcilio Ficino e il Ritorno di Platone*, Padua 1986, 509-31.

————, 'Neoplatonism, the Greek commentators and Renaissance Neoplatonism', in D.J. O'Meara (ed.), *Neoplatonism and Christian Thought*, Norfolk 1982, 169-84, 264-82.

————, 'Albert the Great and the *Studio Patavino* in the late fifteenth and early sixteenth centuries', in J.A. Weisheipl (ed.), *Albertus Magnus and the Sciences: Commemorative Essays*, Toronto 1980, 537-63.

Meyer, H., *Das Corollarium de Tempore des Simplikios und die Aporien des Aristoteles zur Zeit* (=*Monographien zur Naturphilosophie* 8), Meisenheim am Glan 1969.

Mueller, I., 'Aristotle and Simplicius on mathematical infinity', *Proceedings of the World Congress on Aristotle*, vol. 1, Athens 1981, 179-82.

O'Brien, D., *Democritus Weight and Size: an exercise in the reconstruction of early Greek philosophy*, Paris 1981.

————, *Empedocles' Cosmic Cycles*, Cambridge 1969.

Perry, B.M., *Simplicius as a Source of and Interpreter of Parmenides*, Diss., Univ. of Washington, Seattle 1983.

Porawski, R., 'Le fragment du traité d'Aristote de Democrito', *Eos* 71, 1983, 227-81.

Praechter, K., 'Simplikios', *RE* 3 A, 1: (zweite Reihe), 1927, cols 204-13.

Prato, G., 'Contributi dei Mss Marc. Gr. 214 e 227 al testo delle *Fisica* di Aristotele', *Atti dell'Istituto Veneto di Scienze, Lettere ed Arti, Classe di Scienze Morali e Lettere* 132, 1973-4, 105-22.

Ramnoux, C., 'La récupération d'Anaxagore', *Archives de Philosophie* 43, 1980, 75-98.

Roesler, W., '*Omou chrêmata panta ên*', *Hermes* 99, 1971, 246-8.

Rudberg, G., 'In Commentaria Aristotelea fontesque eorum annotationes', Miscellanea G. Mercati, Studi e Testi CXXIV, Città del Vaticano, *Biblioteca Apostolica*, 1946, vol. 4: *Letteratura Classica e Umanistica*, 48-57.

Rudio, F., 'Zur Rehabilitation des Simplicius', *Bibliotheca Mathematica* 3.4, 1903, 13-18.

——————, 'Der Bericht des Simplicius über die Quadraturen des Antiphon und des Hippocrates, *Bibliotheca Mathematica* 3.3, 1902, 7-62.

Ruelle, C.E., 'Clepsydre ou hydraule? (Simpl. in Aristotelis *Phys.* 160v)', *Revue de Philologie* 21, 1897, 110-11.

Sabra, A.I., 'Simplicius' proof of Euclid's parallels postulate', *Journal of the Warburg and Courtauld Institutes* 32, 1969, 1-24.

Sambursky, S., *The Concept of Time in Late Neoplatonism*, Jerusalem 1981.

Sambursky, S., and Pines, S., *The Concept of Place in Late Neoplatonism*, Jerusalem 1982.

Schmidt, W., 'Zu der Bericht des Simplicius über die Möndchen des Hippocrates', *Bibliotheca Mathematica* 3.4, 1903, 118-26.

Sodano, A.R., 'Una polemica di Aristotele sulla concezione platonica della materia', *Rivista Critica di Storia della Filosofia* 18, 1963, 77-88.

Solmsen, F., 'The tradition about Zeno of Elea re-examined', *Phronesis* 16, 1971, 116-41.

Sonderegger, E., *Simplikios. Über die Zeit: Ein Kommentar zum Corollarium de Tempore*, Göttingen 1982.

Sorabji, R., *Matter, Space and Motion*, London and Ithaca, N.Y. 1988.

——————, *Time, Creation and the Continuum*, London and Ithaca, N.Y. 1983.

Tannery, P., 'In Simplicii de Antiphonte et Hippocrate excerpta, pp. 54-59 Diels' in H. Diels (ed.), *Simplicius in Aristotelis Physicorum libros quattuor priores*, Berlin 1882. xxvi-xxxi.

Tardieu, M. *Coutumes nautiques mésopotamiennes chez Simplicius*, in preparation

——————, 'Les calendriers en usage à Ḥarrān d'après les sources arabes et le commentaire de Simplicius à la *Physique* d'Aristote', in I. Hadot (ed.), *Simplicius, sa vie, son oeuvre, sa survie*, Berlin 1987, 40-57.

——————, 'Sabiens Coraniques et "Sabiens de Ḥarrān" ', *Journal Asiatique* 274, 1986, 1-44.

Todd, R.B., 'Infinite body and infinite void. Epicurean physics and peripatetic polemic', *Liverpool Classical Monthly* 7, 1982, 82-4.

Tsouyopoulos, N., 'Die Entstehung physikalischer Terminologie aus der neoplatonischen Metaphysik', *Archiv für Begriffsgeschichte* 13, 1969, 7-33.

Vamvoukakis, N., 'Simplicius, commentateur représentatif d'Aristote dans le néoplatonism tardif', *Proceedings of the World Congress on Aristotle*, vol. 1, Athens 1981, 205.

Vasoli, E., 'Simplicius', *Enciclopedia di Filosofia*, vol. 5, col. 1386.

Verbeke, G., 'Ort und Raum nach Aristoteles und Simplicius. Eine philosophische topologie', in J. von Irmscher and Reimar Mueller (eds), *Aristoteles als Wissenschaftstheoretiker*, Berlin 1983, 113-22.

————, 'Some later Neoplatonic views on divine creation and the eternity of the world', in D.J. O'Meara (ed.), *Neoplatonism and Christian Thought*, Norfolk 1982, 45-53.

————, 'Simplicius', *Dictionary of Scientific Biography* 12, 1975, 440-3.

Wieland, W., 'Die Ewigheit der Weit (der Streit zwischen Joannes Philoponus und Simplicius)' in D. Heinrich, W. Schulz, K.H. Volkmann-Schluck (eds), *Die Gegenwart der Griechen im neueren Denken, Festschrift für H.-G. Gadamer zu 60 Gerburtstag*, Tübingen 1960, 290-316.

Zahlfleisch, J., 'Einige Bemerkungen zur Corollarien des Simplicius in seinem Kommentar zu Aristoteles *Physik*', *Archiv für Geschichte der Philosophie* 15, n.s. 8, 1902, 186-213.

————, 'Variae lectiones zur *Physik* E-Z des Aristoteles bei Simplikios (801,14-861,28)', *Philologus* 59, 1900, 64-89.

————, 'Die Polemik des Simplicius gegen Aristoteles *Physik* 4, 1-15 über den Raum dargestelt', *Archiv für Geschichte der Philosophie* 10, 1897, 85-109.

Philosophers cited by Simplicius

Alexander (of Aphrodisias). Third-century A.D. Peripatetic commentator on the works of Aristotle; his commentary on the *Physics* is lost.

Andronicus. First-century B.C. Peripatetic, and head of the school; wrote a treatise on Aristotle's works, and arranged them in their modern order.

Aristotle. Fourth-century B.C. philosopher from the city of Stagira, who founded the Peripatetic school in Athens.

Aspasius. A Peripatetic commentator on Aristotle, flourished in the first half of the second century A.D.; his commentary on the *Physics* is lost.

Damas. Author of a *Life of Eudemus*; nothing further is known of him.

Democritus. Fifth-century B.C. materialist philosopher, originally from the northern Greek city of Abdera; chief originator of the ancient atomic theory.

Diodorus Cronus. Flourished *c.* 300 B.C.; elaborated dialectical arguments in the tradition of Zeno, and developed a version of atomism.

Diogenes the Cynic. Fourth-century B.C. ethical thinker, famous for his defiance of convention, and for his pointed, iconoclastic wit.

Epicurus. Athenian philosopher of the third/second centuries B.C., who refined Democritean atomism, partly in response to Aristotle's criticisms.

Eudemus. Pupil of Aristotle, from the island of Rhodes; flourished in the second half of the fourth century B.C.; compiled a paraphrase of Aristotle's *Physics*, among other works.

Leucippus. Fifth-century B.C. philosopher, associated with Democritus as the inventor of the atomic theory; little is known of his life and specific teachings.

Melissus. Fifth-century B.C. philosopher from the island of Samos; a critic of the phenomena of change in the tradition of Parmenides and Zeno.

Peripatetics. Adherents of the doctrines of Aristotle.

Plato. Athenian philospher of the fifth/fourth centuries B.C., and teacher of Aristotle; originator of the theory of Forms or Ideas.

Strato of Lampsacus. Pupil of Theophrastus and his successor as head of the Aristotelian school (third century B.C.); developed the Peripatetic theory of matter and space.

Themistius. Fourth-century A.D. orator and philosopher in Constantinople, author of paraphrases of the works of Aristotle (his paraphrase of the *Physics* is extant).

Theophrastus. Pupil of Aristotle, originally from Eresus on the island of Lesbos, flourished in the fourth/third centuries B.C.; succeeded Aristotle as head of the Peripatetic school.

Timon. Third-century B.C. sceptical philosopher, and author of witty lampoons on the dogmatic thinkers.

Zeno. Fifth-century B.C. philosopher, originally from Elea, who propounded a series of famous paradoxes calling into question the phenomena of motion and change.

Indexes

Greek-English Index

elleima, shortfall, 973,10
elleipein, fall short, 949,10.11.13
  bis.14.15.17.21.24; 950,10;
  973,10.11
  be short of (something), 975,8
emphainein, exhibit, 974,28
enantion, contrary, 923,14; 942,3;
  996,8.29; 1019,15.31; 1021,24; 1030
  passim
enantiôsis, contrariety, 1030,18
enargeia, self-evidence, 1012,26
enargês, self-evident, obvious, 933,7 et
  passim
enargôs, manifestly, 962,20 et passim
endeiknunai, indicate, 944,13; 954,26;
  983,26; 1019,22.24; 1025,8
endeiknusthai, show, indicate, 1019,22;
  1025,8
endekhesthai, be possible, 924,4 et
  passim
energeia, activity, 923,18
  actuality, 923,18 et passim
  energeiâi, (dat.) in actuality, 928,26 et
  passim
energein, act, 942,3
enestôs, present, 955,5 et passim
enistasthai, object, raise objections,
  999,3; 1034,20, et passim
  be in objection to, 925,18
enkosmios, in the world, 966,1
enkuklios, revolving, 1030,15
ennoein, note, consider, think up, 942,2;
  1005,21; 1009,22; 1012,14; 1016,3
ennoia, notion, 925,7; 1008,31
enstasis, objection, 930,16 et passim
entelekheiâi, (dat.) actually, 948,14
enudron, water-creature, 961,17
enulos, (adj.) material, 966,9
enuparkhein, inhere, 955,15; 997,22;
  1004,33; 1025,3.5.6.7.27
epagein, add, 926,29 et passim
  draw (a consequence), 934,11 et
  passim
epanodos, return (n.), 1031,26
epeisienai, enter, 965,4
epekeinos, yonder (side), 955,29
epekhein, hold, 1016,26
ephaptesthai, touch upon, 927,7
epharmozein, be superimposed,
  927,3.7.9; 931,23
  fit, 964,27.28
ephexês, next, 924,25 et passim
  and so on, 923,4
  consecutive, 924,22 et passim
ephistanai, pay attention, attend,
  936,25; 945,6 et passim
  remark, 951,30

ephodos, method, 936,13; 954,17
epiballein, address, 946,26; 996,7
epideiknunai, show, 945,5; 952,3 bis;
  953,10
epidromê, summary, 1031,32
epigraphein, label, 923,3.5; 933,14
epikheirêma, argument, 928,5.28 et
  passim
epikheirêsis, reasoning, 930,34; 972,1;
  1032,33
epikrinein, decide, 979,6
epilambanein, take over, 1028,11
epilambanesthai, attain, 1015,19;
  1016,14
epileipein, fail, 989,8; 993,31.32
epinoeisthai, think of, 946,4
epipedon, plane, 953,13.14; 964,8
epiphaneia, surface, 925,5 et passim
epipherein, conclude, 981,20; 989,18
epiphora, result, 929,30
epitasis, intensification, 999,31
epiteinein, intensify, 999,29.31
epitelein, accomplish, complete, 967,21;
  991,14.16.18.20.31; 984,14; 986,31;
  1005,24
epithumia, appetite, 964,31
êremein, be at rest, 923,13 et passim
êremêsis, resting, 1001,20.21.24.26;
  1002,17; 1023,31
êremia, rest, 923,14 et passim
êremizein, bring to rest, 1001,24.28;
  1002,17.18.21.22.23; 1006,20.22;
  1007,5.7 tris; 1008,28; 1033,24
eskhaton, extreme, 925,4 et passim
eskhatos, last, extreme, 974,3; 1017,5
eulogôs, logically, 932,20
euparadektos, readily acceptable,
  1023,1
eupathês, easily affected, 968,24
euphradês, clever, 968,30
euthus, straight, straightway, 946,18;
  1008,2 et passim
exaiphnês, sudden, 982,6
exairetos, exceptional, 957,21
exaptein, fasten, 966,4
exartan, attach, 965,23; 966,1
exêgeisthai, explain, 936,22.30; 982,7
exêgêtês, commentator, 946,25; 991,28
exêgêtikos, explanatory, 988,10
exisôthein, make equal, 949,22
existasthai, depart from, 1024,3

gegonos, past, 955,19 et passim
geloios, ridiculous, 969,21
genesis, coming to be, 946,21 et passim
genos, kind, 930,1.4; 982,14
gignesthai, occur, be made, become,

arise, turn, up 939,7 et passim
come to be, 995,25 et passim
be produced, 1028,4
gnômê, thought, 991,29
gnôrimos, familiar, 975,23; 998,5
grammê, line, 925,5 et passim
graphê, reading, 936,4.29; 949,30;
  958,24; 965,17
  picture, 965,17
graphein, write, draw, 923,10; 1026,6 et
  passim
gumnazein, exercise, 939,28; 1012,23

hama, simultaneously, at the same
  time, 933,14; 958,13 et passim
hapaxapas, all together, 967,17
haphê, contact, 997,32; 998,16.19
haplôs, simply, 929,5 et passim
hapsis, vault (of heaven), 966,11
haptesthai, touch, 924,22 et passim
harmozein, fit, 980,17; 996,32; 1029,4;
  1030,20; 1033,19
hêgoumenon, antecedent, 932,24;
  953,22; 970,14.15; 997,19
hepomenon, consequent, 929,29 et
  passim
  (in) consequence, 934,11; 978,21
hidruein, settle, 942,14
hiesthai, head (in a direction), 1002,22
histasthai, halt, 1001,25 et passim
  (pf.) stand still, 1016,5
hodeuein, travel, 963,30
hodos, route, 1007,6
holos, whole, 926,14 et passim
holôs, in general, 926,15.26 et passim
  at all, 969,7 et passim
homalôs, evenly, 941,22; 948,31; 974,21;
  975,29; 1002,20
homogenês, kin, akin,
  928,8.10.11.14.17.19.29; 929,10.15
homoios, similar, 986,13; 989,8; 1013,24
homoiôs, similarly, 924,26 et passim
  uniformly, 969,30
homoiotês, similarity, 932,14; 948,19;
  989,10; 1032,5
homônomia, equivocality, 947,32
homophuês, of like nature, 966,9
homophuôs, in a naturally similar way
  944,13
homotakhês, of a similar speed, 993,1
homotakhôs, at a similar speed, 992,24
hopostêmorios, in proportion, 973,24
horasis, seeing, 961,16; 965,13
horismos, definition, 938,13 et passim
  boundedness, 1030,19
horizein, determine, define, mark off,
  bound, 942,10; 1015,24; 1016,27;

1030,2 et passim
horizesthai, define, 1025,13
horman, start, set out, 1025,2; 1034,20
horos, definition; boundary, 965,6
hugiês, sound (of an argument), 930,11;
  963,11; 997,6; 978,4
hupantan, address, 966,27
huparkhein, belong, pertain, 958,19
  A *huparkhei* B = B has A,
  1006,12.13.18; 1007,15; 1008,24
  be a property of, 941,22
  exist, 955,24
huparchon, property, 1033,27.29.30
huperbainein, skip over, 947,13
huperballein, exceed, 949,10.11.13.14
  bis.21; 950,10; 973,5.10
huperblêma, excess, 973,7.9.10
huperekhein, exceed, 1015,7
huperokhê, excess, 954,20; 1029,13
huphistasthai, consist, 925,13
hupokeisthai, underlie, 988,18; 989,22
  be hypothesized, 952,7
  be prescribed, 1014,23.24
hypokeimenon, subject, 1030,3 bis
hupoleipesthai, be remaining, 1014,28
hupomimnêskein, mention, 941,17;
  961,12; 994,29
  remind, 1009,11
  recall, 1008,6
hupomnêsis, reminder, 930,9
huponoia, consideration, 934,11
hupopiptein, be subject to, 995,22
  come under, 1022,21
hupostasis, existence, 928,22; 984,19
hupotattein, be classified under, 982,15
hupothesis, hypothesis, 934,18; 950,13;
  954,4; 964,14; 980,9; 984,31
hupothetikos, hypothetical, 932,15;
  953,19; 961,24
hupotithenai, hypothesize, 936,30 et
  passim

idikos, individual, 993,1
idios, individual, 925,7; 935,10; 948,30;
  973,30
  own (adj.), 978,12
  personal, 1014,8
idiôs, specifically, 924,16
isarithmos, numerically equal, 943,33
isazein, equal (vb.), 973,25
isêmerinos, equinoctial, 1026,7
isokhronios, contemporaneous, 960,2
isoonkos, of equal volume, 1016,23
isopleuros, equilateral, 982,16
isos, equal, 938,12 et passim
isotakheia, equal speed, 1019,23
isotakhês, equal in speed, of equal (or

uniform) speed, 936,31; 1016,11 et
passim
isotakhôs, at the same (or a uniform)
speed, 936,7.10.18; 940,25 et
passim
isotês, equality, 1019,23

kakunein, corrupt, 942,6
kanonizein, prescribe for, 980,23
katalambanein, overtake, 987,3 et
passim
katalêgein, end up, 925,9
kataleipein, leave over, 993,31
katametrein, measure off, 949,8 et
passim
katanankizein, place under necessity,
941,27
katanoêsis, observation, 1027,14
katantan, arrive, 1015,21
kataphasis, affirmation, 1022,4
kataskeuazein, establish, 932,23; 1014,4
katatattein, set forth, 935,10
katêgorein, predicate, 993,25; 994,2;
997,31
katêgoria, category, 928,22.23
katêgorikos, categorical, 962,17
katekhein, occupy, 1010,23.28; 1012,13;
1028,8
kathairein, reduce, 995,17; 996,18
kathairesis, reduction, 996,23
kathareuein, be pure, 942,11
kath'hauto (kath'heauto), in its own
right, 958,13 et passim
kath'holon heauto, in its own right as a
whole, 1023,17
katholikos, general, 980,17; 999,5;
1026,29
katholou, in general, 925,22 et passim
keisthai, be supposed, be posited,
934,15.20 et passim
be posited, 934,20
keimenon, supposition, 943,18
kenon, empty (space), 928,30 et passim;
(adv.) 984,19
khôrismos, separateness, 965,5
khôristos, separable, 965,25; 1024,27
khôrizein, separate, 926,23 et passim
khreia, function, 1006,5; 1008,21
khrêsimos, useful, 1011,16
kineisthai, move (intrans.), moving,
932,23 et passim
kinêma, move (n.), 934,12; 957,18;
986,17; 1027,19.22
kinein, move (trans.), set in motion,
964,22; 1012,18
kinêsis, motion, 923,14 et passim
kinêtos, movable, 923,18 bis

koinoun, share, 967,18
koinos, general, 980,27
koinôs, in general, 933,3
kosmos, world, 966,2
krinein, decide, 964,25
krisis, decision, 965,2
ktasthai, (perf.) possess, 969,18
kubos, cube, 1016,24
kuklikos, circular, 946,20
kuklikôs, circularly, 1023,6
kuklophorêtikos, revolving, 941,22
kuklophoria, revolution, 946,16
kuriôs, in a strict sense, 948,23 et
passim

lambanein, take, assume, adopt, 939,22;
956,18 et passim
draw (a consequence), 956,9
legomenon, proposition, statement,
984,24; 1008,23
lêgon, consequent, 970,15
leptos, fine, 966,8
lêptos, able to be taken, 986,30
lexis, word, statement, passage, 929,18;
994,27; 997,30
logikos, rational, 965,9
logos, reasoning, argument, 932,1;
934,31 et passim
ratio, 939,19 et passim
(ana) logon, corresponding, analogous,
1023,21; 940,2
luein, solve, resolve, 930,16 et passim
lusis, solution, 948,8
lutikos, in refutation, 958,15

malthakôs, weakly, 941,24
marturia, evidence, 969,4
matên, in vain, 1030,26.27
mê einai, non-existence, 947,5 et passim
megethos, magnitude, 924,4 et passim
meiôsis, decrease, 968,2.5; 975,17.22;
977,20; 989,9; 1030,10
mêkos, length, 927,28 et passim
mellon, future, 955,14 et passim
mêpote, but, now (introducing a comment by S.), 929,4 et passim
merikos, particular, 946,18
merismos, being partitionable, 997,7
meristos, partitionable, 933,29 et
passim
merizein, partition, 937,21; 942,21;
974,7; 987,7
meros, part, 925,6 et passim
(para) meros, by turns, 999,29
mesos, intermediate, middle, 955,1;
1016,26 et passim
metabainein, switch, 933,8

pass to, 1004,20; 1026,18
metaballein, change, 938,30 et passim
metabasis, passage, 1035,8
metabatikôs, in passing about, 1030,14
metabolê, change (n.), 924,24 et passim
metagein, carry over, 941,1
metakheirizesthai, handle, 1001,19
metalambanein (A *eis* B), substitute (B for A), 929,32
  (A *ek* B), substitute (A for B)
  take up in turn, 1023,29
metalêpsis, alternation, 943,25
  transfer, 979,21.23
metallattein, undergo a change, 992,17
metapiptein, migrate, 959,23
metaxu, between, in between, 928,2 et passim
  meanwhile, 1019,11
metekhein, partake, 946,23
meterkhesthai, shift, 988,7
methodos, method, 1030,19
metrêsis, measurement, 949,10
metron, measure (n.), 945,17; 946,29
metroun, measure, 945,17.18.20; 946,6.18; 950,11; 951,19
moira, section, 942,16
monas, unit, 926,19
morion, portion, 926,16 et passim

neuein, (perf.) face, 1017,14
noein, think, 965,11.16; 973,28
noêsis, thinking, 965,12
nomizein, believe, 923,16 et passim
nous, mind, 965,4
nun, now, a now (i.e. instant), 926,19 et passim
  here, 924,17 et passim

ôdinein, be in labour with, 1031,29
okhêma, vehicle (for the soul), 964,19; 965,22.23
oikeios, proper, 928,10
  own, 941,26; 964,14; 966,2; 1028,5; 1030,11
on, existing, 944,28 et passim
onkos, volume, 1017,1.7; 1028,5
onoma, name, term, epithet, 958,19; 979,21; 1005,20; 1014,8
orexis, desire, 964,31
orgê, feeling of anger, 965,1
orthos, right, 976,30; (adv.) 929,4; 967,30
  (of angles), 982,16
ostreôdês, oyster-like (of the vehicle of the soul), 966,5
ouranios, heavenly, 942,4.9

ousia, essence, 965,24.25
pakhus, crass, 966,9
panteleios, completely perfect, 965,4
pantôs, strictly; altogether, 930,25 et passim
  invariably, 931,17 et passim
paradeigma, example, 967,20.24; 1024,23
paradoxologein, speak paradoxically, 1006,16
paradoxon, paradox, 1009,21
paradoxoteros, more paradoxical, 960,24; 1011,27
paragignesthai, arrive, 1014,13
paragein, bring in, 942,18 et passim
parakeisthai, be adjacent, 981,7
paralambanein, take, 933,7; 956,22; 967,21; 1000,17; 1014,6.10
paraleipein, leave out, 1027,4
parallêlos, parallel, 938,10
  (*ek*) *parallêlou*, redundantly, 1024,27
paralogismos, fallacy, 1018,13; 1019,24; 1020,1
paralogizesthai, reason falsely, 1011,9; 1024,14
parangellein, invoke, 1021,12
paratithesthai, cite, 967,20; 991,28
paraxein, follow closely, 924,18
parêkôn, past, 957,4 et passim
parelêluthos, past, 955,13 et passim
parelkein, mislead, 969,12
paremballein, insert, 1025,15
parerkhesthai, pass by, 1004,15
parienai, omit, neglect, 929,30; 932,28; 933,6; 972,26; 975,23
paristan, exhibit, 932,15
parousia, presence, 976,12.13.15.16.21.24.25; 977,10.11.12
paskhein, undergo, 989,30
pathos, state (n.), 976,11; 981,1
pauesthai, stop (intrans.), 981,12; 984,9; 992,15.21; 994,10
pêgnuein, freeze, 966,20; 968,22; 969,1
pelazein, be adjacent, 927,10
peperasmenos, finite, 924,8 et passim
perainein, delimit, 950,26; 999,18.22
peras, limit (n.), 926,13 et passim
peratoun, limit, 956,13
periaptein, hang around, 964,19
peridineisthai, spin round, 1024,4
periekhein, surround, 955,10.11; 1023,12
perigegonotôs, for good measure, 1023,16
perigraphein, comprise, 931,7
(*ek*) *periousias*, for good measure,

pros to, relevant, 966,30
prostragôidein, add in tragic style, 1015,8
protasis, promise, 1021,31
protithesthai, propose, 925,22 et passim
prôtos, first, prior, primary, 923,10; 968,14; 982,8 et passim
prôtôs, primarily, 955,7 et passim
pseudos, false, 939,5; 947,1
   pseudos, falsity (n.), 1015,31; 1016,8
psukhikos, of the soul, 964,27
psukhoun, animate (vb.), 966,3

rhêsis, passage, 938,6; 949,30
rhêton, statement, 965,13

saphêneia, clarification, 1031,31
sêmainein, mean, 929,18 et passim
sêmainomenon, meaning, 998,18
sêmeion, point, 929,15 et passim
   sign, 968,23
skhêma, (logical) figure, 926,6; 928,10; 962,17; 1001,8 et passim
skhetikôs, relatively, 965,27
skopos, aim, 1022,14; 1024,18
sôma, body, 925,5 et passim
sômatikos, corporeal, 964,24; 965,1.19
sophisma, paradox, 947,32; 1012,26
sophismos, sophistry, 1020,11
sophistikôs, sophistically, 1020,10; 1024,16
sophizesthai, quibble, 1022,13
sphairikôs, spherically, 1023,6.8
stasis, stop (n.), halt, 952,10; 1001,25 et passim
sterêsis, privation, 923,14; 1006,26; 1007,3; 1009,8
steresthai, be deprived, 1006,27.30.31.32.33; 1009,8; 1027,14
stigmê, point, 926,1 et passim
stoikheion, letter, 923,4.6; 939,27
sungenês, akin, kin, 928,11.18
sunkeisthai, be composed (of), 924,19 et passim
sunkhrasthai, make joint use, 996,12
sullogismos, syllogism, 932,16.23; 962,16; 1001,8; 1015,27
sullogizesthai, syllogize, 928,11; 932,26; 936,2; 953,19; 956,17; 1011,22; 1016,5; 1026,32
sumbainein, result (vb.), happen that, 931,15; 960,1; 1012,12 et passim
   (*kata*) *sumbebêkos*, incidentally, 964,7 et passim
sumparalambanein, take (X) along with (Y), 1005,17

sumparateinein, coextend (alongside), 991,14.20; 1006,12
sumpathês, sympathetic, in accord, 925,19; 974,6
sumperainein, delimit, co-limit, 925,19; 1003,7
sumperainesthai, conclude, 973,15; 983,14; 987,28; 996,13; 1003,7; 1031,7
sumperasma, conclusion, 929,30; 996,5
sunagein, infer, 928,14 et passim
   yield, 1034,17
sunagôgê, inference, 926,6; 956,6
sunaleiphesthai, coalesce, 931,17
sunalêtheuein, be true together with, 1021,14.15
sunanairein, co-eliminate, 1006,10
sunapartizein, co-complete exactly, 1001,16; 1005,9
sunapodeiknunai, demonstrate simultaneously, 925,7; 933,3; 944,11
sunapokathistasthai, come back around together, 1026,4.10
sunaptein, attach, 924,18; 932,29
sunarithmein, count along with, 988,23
sundapanan, co-exhaust, 1003,4
sundiairein, co-divide, 936,10.23 et passim
sundiiistasthai, be co-extensive, 1006,8
sunekheia, continuity, 957,13.18
sunekhês, continuous, 924,19.21 et passim; (adv.) 935,5 et passim
sunekhizesthai, be in continuity, 926,27; 930,25.26
sunekteinein, coextend, 1006,7
(hôs) sunelonti (eipein), in brief, in short, 932,4; 944,7
sunêmmenon, conditional premise, 929,28 et passim
sunengismos, approach, 1022,2
sunistanai, support, 945,2
sunistasthai, be constructed, 925,23; 932,10; 934,27
sunônumos, denoted by the same name 957,8.11
sunthesis, composition, 1028,28
suntrekhein, run together, 928,8

takhos, speed, 939,19; 942,21; 1025,33
takhutês, speed, 1014,9
tautotês, same condition, 942,13
taxis, order, position, 923,3.4.6; 924,17.20; 1000,9
teinein, extend, 1005,23
tekhnologein, systematically discuss, 1033,27
tekmairesthai, judge, 1008,31

teleioun, complete (vb.), 997,14
teleios, full, perfect, 942,8; 966,11;
 1030,11; 1033,26
teleutaion, ending, 987,2
teleutaios, last, final, 924,6; 946,12 et
 passim
teleutan, end (vb.), 1017,3
teleutê, end (n.), 924,3; 955,23; 966,3;
 996,3
telikos, final, 931,18
telos, end (n.), 938,5; 955,15 et passim
thaumastos, amazing, 986,6; 1009,23;
 1011,15; 1023,18
theios, divine, 942,5.9.11; 952,9; 965,4
theôrêma, theorem, 1005,18; 1009,21;
 1011,15; 1012,3.9; 1030,1
theôrein, speculate, 964,25
theôria, speculation, 965,2
thesis, position, 1017,16; 1023,12.29
thespesios, extraordinary, 1011,11
tithenai, posit, set, pose, put down,
 924,1; 933,3; 934,11; 975,20
 form, 932,19
tithesthai, make, 987,20

tmêma, segment, 958,10; 963,10.11;
 983,14; 1003,6
tode ti, particular thing, 982,8
toionde, (particular) sort, 982,8
tomê, cutting, cut, 925,11.12; 959,27 et
 passim
topikos, local, 967,21; 1005,26.31;
 1023,21; 1028,5.15; (adv.)
 965,26.28; 967,8
topôi, spatially, 926,23; 927,18.19 et
 passim
topos, place, passage, 928,9; 968,30 et
 passim
 (*kata*) *topon*, locally, 967,5
tosonde, (particular) amount, 982,9
 this much, 989,11
tropos, way, character, form, 927,1;
 945,4; 961,10 et passim

zêtein, inquire, investigate, 930,10,
 956,21 et passim
zêtêsis, investigation, 1024,8
zôidiakon, zodiac, 942,16
zôidiakos, zodiacal, 1026,7

English-Greek Glossary

abate: *anienai*
abatement: *anesis*
ability: *dunamis*
able: *dunatos*
absurd: *atopos*
acceptable, readily: *euparadektos*
accomplish: *anuein, dinuein, apitelein*
accord, in: *sumpathês*
accurate: *akribês*
act: *energein*
activity: *energeia*
actuality: *energeia*
actuality, in: *energeiâi* (dat.)
actually: *entelekheiâi* (dat.)
add: *epagein*
added, be: *prokeisthai*
addition: *prosthêkê, prosthesis*
address: *epiballein, hupantan*
adjacent, be: *parakeisthai, pelazein*
admit: *prosdekhesthai*
adopt: *lambanein*
advance: *proerkhesthai*
affected, easily: *eupathês*
affirmation: *kataphasis*
ahead, be: *proagein*
aim: *skopos*
air: *aêr*
akin: *homogenês, sungenês*

all at once (of change): *athroos*
all together: *hapaxapas*
alter: *ameibein*
alternation: *metalêpsis*
altogether: *pantôs*
amazing: *thaumastos*
ambivalent: *amphibolos*
amount (particular): *tosonde*
analogous: (*ana*) *logon*
analogous, be: *analogein*
and so on: *ephexês*
anger, feeling of: *orgê*
animate (vb.): *psukhoun*
another place, in: *allakhêi*
antecedent: *hêgoumenon*
antecedently: *proêgoumenôs*
antithetical opposition: *antithesis*
appearance: *phainomenon*
appetite: *epithumia*
approach: *sunengismos*
argue briefly: *brakhulogein*
argue: *dialegesthai*
argument: *epikheirêma, logos*
arise: *gignesthai*
arrive: *katantan, paragignesthai*
articulate: *diarthroun*
ascent: *anabasis*
assign: *apodidonai*

assume in addition: *proslambanein*
assume in advance: *prolambanein*
assume: *lambanein*
assumption in advance: *proeilêmmenon*
atemporally: *akhronôs:*
atomic, atom: *atomos*
attach: *exartan, sunaptein*
attain: *epilambanesthai*
attend, pay attention: *ephistanai*
avoid: *ekklinein*
axiom: *axiôma*

basic: *anankaios*
bear: *pheresthai*
become: *gignesthai*
before, be: *prokeisthai*
begin: *arkhesthai*
beginning: *arkhê*
beginningless: *anarkhos*
believe: *nomizein*
belong: *huparkhein*
bend back (intrans.): *anakamptein*
between, in between: *metaxu*
body: *sôma*
border (on): *ekhesthai*
both: *allos kai allos*
bound: *horizein*
boundary: *horos*
boundedness: *horismos*
breadth: *platos*
brief, in: *(hôs) sunelonti (eipein)*
bring in: *paragein*
bring: *pheresthai*
broad (sense), in a: *(en) platei*
but (introducing a comment by S.):
    mêpote

can: *dunasthai*
can have (an attribute): *dektikos*
carry over: *metagein*
categorical: *katêgorikos*
category: *katêgoria*
cause: *aitia*
change (n.): *metabolê*
change: *metaballein*
change, undergo a: *metallattein*
character: *tropos*
circular (proof): *diallêlos*
circular: *kuklikos*
circularly: *kuklikôs*
circumference: *periphereia*
cite: *paratithesthai*
clarification: *saphêneia*
classified under, be: *hupotattein*
clever: *euphradês*
co-complete exactly: *sunapartizein*
co-divide: *sundiairein*

co-eliminate: *sunanairein*
co-exhaust: *sundapanan*
co-extensive, be: *sundiistasthai*
co-limit: *sumperainein*
coalesce: *sunaleiphesthai*
co-extend (alongside): *sumparateinein*
co-extend: *sunekteinein*
come back around together:
    sunapokathistasthai
come previously to be: *progignesthai*
come to be: *gignesthai*
come under: *hupopiptein*
coming to be: *genesis*
commentator: *exêgêtês*
complete (vb.): *teleioun, epitelein*
complete exactly: *apartizein*
composed (of), be: *sunkeisthai*
composition: *sunthesis*
comprise: *perigraphein*
conclude: *epipherein, sumperainesthai*
conclusion: *sumperasma*
condition: *diathesis*
condition, same: *tautotês*
conditional premise: *sunêmmenon*
confirm: *bebaiousthai, pistoun*
confirmation: *pistis*
conform: *akolouthein*
conformity: *akolouthia*
consecutive: *ephexês*
consequence: *akolouthia, hepomenon*
consequence, in: *akolouthôs*
consequent: *hepomenon, lêgon*
consequently: *ara*
consider: *dianoeisthai, ennoein*
consideration: *huponoia*
consist: *huphistasthai*
constituted, be so (i.e. be of such a
    nature): *phuein* (pf.)
constrain: *anankazein*
constructed, be: *sunistasthai*
contact: *haphê*
contemporaneous: *isokhronios*
continuity: *sunekheia*
continuity, be in: *sunekhizesthai*
continuous: *sunekhês*
contradiction: *antiphasis*
contradictorily: *antiphatikôs*
contradictory: *antiphatikê*
contradictory pair: *antiphasis*
contrariety: *enantiôsis*
contrary: *enantion*
converse: *antistrophos*
conversely: *anapalin*
conversion (logical): *antistrophê*
convert: *antistrephein*
convertible, be: *antistrephein*
corollary: *porisma*

corporeal: *sômatikos*
correctly: *alêthôs*
correlative: *pros ti*
corresponding: *(ana) logon*
corrupt: *kakunein*
count along with: *sunarithmein*
count: *arithmein*
crass: *pakhus*
cube: *kubos*
cut, cutting: *tomê*

decide: *diairein, epikrinein, krinein*
decision: *krisis*
decrease: *meiôsis*
define: *horizein, horizesthai*
definition: *horismos, horos*
deliberate: *bouleuesthai*
delimit: *perainein, sumperainein*
demonstrate: *apodeiknunai*
demonstrate in addition:
    prosapodeiknunai
demonstrate in advance:
    proapodeiknunai
demonstrate simultaneously:
    sunapodeiknunai
demonstration: *apodeixis*
denial: *anairesis*
deny: *anairein*
depart from: *existasthai*
departure: *ekstasis*
depend: *artasthai*
deprived, be: *steresthai*
desire: *orexis*
detach: *apospan*
determine: *horizein*
dichotomy: *dikhotomia*
differ: *diapherein*
difference: *diaphora*
difference, make a: *diapherein*
different: *diaphoros*
different kind, of a: *anomogenês*
difficult to confront: *dusantibleptos*
difficulty: *duskolia*
diminution: *phthisis*
discuss: *dialegesthai*
discuss, systematically: *tekhnologein*
disjunction: *diairetikon*
dissimilar type, of: *anomoeidês*
dissolve: *dialuein*
distinction: *diorismos*
distinguish: *antidiairein, diorizein*
distinguish in advance: *prodiorizesthai*
divergence: *diastolê*
divide: *diairein*
divine: *theios*
divisibility: *(to) diaireton*
divisible: *diairetos*

division: *diairesis*
divisor: *diairetikon*
draw (a consequence): *epagein,*
    lambanein
drawn (in an illustration): *ektithesthai*
draw a distinction: *diorizesthai*
draw: *graphein*

eclipse, be in: *ekleipein*
embrace: *ekhein*
emerge: *ekbainein*
empty (space): *kenon*
end (n.): *teleutê, telos*
end (vb.): *teleutan*
end up: *katalêgein*
ending: *teleutaion*
endless: *ateleutos*
endless(ness): *akatalêktos*
enter: *epeisienai*
epithet: *onoma*
equal (vb.): *isazein*
equal: *isos*
equal, make: *exisôthein*
equality: *isotês*
equilateral: *isopleuros*
equinoctial: *isêmerinos*
equivocality: *homônomia*
essence: *ousia*
establish: *kataskeuazein*
eternal: *aidios*
even (of numbers): *artios*
evenly: *homalôs*
evidence: *marturia*
evident: *phaneros*
exact: *akribês*
exactly: *apêrtismenôs*
example: *paradeigma*
exceed: *huperballein, huperekhein*
exceptional: *exairetos*
excess: *huperblêma, huperokhê*
exercise: *gumnazein*
exhaust: *dapanan*
exhaustive: *anellipês*
exhibit: *emphainein*
exhibit: *paristan*
exist: *huparkhein*
existence: *hupostasis*
existing: *ôn*
explain: *exêgeisthai*
explanatory: *exêgêtikos*
exploit: *apokhrasthai*
expound: *apodidonai*
extend: *teinein*
extent: *diastasis, ektasis*
extraordinary: *thespesios*
extreme (n.): *eskhaton*
extreme: *eskhatos*

likely: *eikos*
limit (n.): *peras*
limit (vb.): *peratoun*
line: *grammê*
local: *topikos*
locally: *(kata) topon*
locomotion: *phora*
logically: *eulogôs*

made, be: *gignesthai*
magnitude: *megethos*
magnitude, without: *amegethês*
maintain: *diaphulattein*
make: *poiein, tithesthai*
manifest: *prophanês*
manifestly: *enargôs*
manuscript: *antigraphos*
mark off: *horizein*
material (adj.): *enulos*
mean: *sêmainein*
meaning: *sêmainomenon*
meanwhile: *metaxu*
measure (n.): *metron*
measure back: *anametrein*
measure (vb.): *metroun*
measure off: *katametrein*
measure, for good: *(ek) periousias, perigegonotôs*
measureless: *ametros*
measurement: *metrêsis*
mention: *hupomimnêskein*
method: *ephodos, methodos*
middle: *mesos*
migrate: *metapiptein*
mind: *nous*
mislead: *parelkein*
motion: *kinêsis, phora*
motionless: *akinêtos*
movable: *kinêtos*
move (intrans.), moving (verbal n.): *kineisthai*
move (n.): *kinêma*
move (trans.), set in motion: *kinein*
move locally: *pheresthai*
move opposite: *antikineisthai*
multiply: *pollaplasiazein*
multitude, in: *plêthei*
mutually implied, be: *antakolouthein*

name: *onoma*
name, denoted by the same: *sunônumos*
naturally similar way, in a: *homophuôs*
nature: *phusis*
nature, of like: *homophuês*
necessary: *anankaios*
necessitate: *anankazein*
necessitate in addition: *prosanankazein*

necessity: *anankê*
necessity, place under: *katanankizein*
negation: *apophasis*
neglect (vb.): *parienai*
newborn: *artigenês*
next: *ephexês*
non-existence: *mê einai*
non-uniform speed, at a: *anisotakhês*
note: *ennoein*
notion: *ennoia*
now (introducing a comment by S.): *mêpote*
now, a now (i.e. instant): *nun*
number: *arithmos*
numerically: *arithmôs*
numerically equal: *isarithmos*

object, raise objections: *enistasthai*
objection: *enstasis*
objection to, be in: *enistasthai*
observation: *katanoêsis*
obvious (in advance): *prodêlos*
obvious: *dêlos*
obvious: *enargês*
occupy (a position): *apoklêrousthai* (pf.)
occupy: *katekhein*
occur: *gignesthai*
omit: *parienai*
one after another: *allos kai allos*
opinion: *doxa*
opposed, be: *antikeisthai*
opposite: *antikeimenos*
order: *taxis*
originally: *(tên) arkhên*
otiose: *perittos*
overshoot: *pleonazein*
overtake: *katalambanein*
own (adj.): *idios, oikeios*
oyster-like (of the vehicle of the soul): *ostreôdês*

paradox: *paradoxon, sophisma*
paradoxical, more: *paradoxoteros*
paradoxically, speak: *paradoxologein*
parallel: *parallêlos*
part: *meros*
partake: *metekhein*
particular: *merikos*
particular thing: *tode ti*
partition: *merizein*
partitionable: *meristos*
partitionable, being: *merismos*
partless: *amerês*
pass by: *parerkhesthai*
pass to: *metabainein*
passage (in a text): *lexis*
passage: *metabasis, rhêsis, topos*

passing about, in: *metabatikês*
past: *gegonos, parêkôn, parelêluthos*
penetrate: *eiskrinesthai*
penetration: *eiskrisis*
perception: *aisthêsis*
perfect: *teleios*
perfect, completely: *panteleios*
perish: *phtheirein*
perishing: *phthora*
personal: *idios*
pertain: *huparkhein*
picture: *graphê*
place: *topos*
plane: *epipedon*
plausibility: *pithanotês*
point: *sêmeion, stigmê*
portion: *morion*
pose: *tithenai*
posit: *tithenai*
posited, be: *keisthai*
position: *taxis, thesis*
possess: *ekhein, ktasthai* (pf.)
possible: *dunatos*
possible, be: *dunasthai, ekhein,*
   endekhesthai
potentially, potential: *dunamei*
power: *dunamis*
power, have the: *dunasthai*
precede: *proêgeisthai, prolambanein*
predicate: *katêgorein*
premise, additional: *proslêpsis*
premise, conditional: *sunêmmenon*
prescribe for: *kanonizein*
prescribed, be: *hupokeisthai*
presence: *parousia*
present (time): *enestôs*
primarily: *prôtôs*
primary: *prôtos*
principle: *arkhê*
prior: *prôtos*
prior, be: *proêgeisthai*
privation: *sterêsis*
problem: *problêma*
procedure: *agôgê*
proceed: *pheresthai, proerkhesthai*
produce: *poiein, poieisthai*
produced, be: *gignesthai*
promise: *protasis*
proof: *deixis*
proof of, in: *deiktikon*
proper: *oikeios*
property: *huparchon*
property of, be a: *huparkhein*
proportion: *analogia*
proportion, in: *hopostêmorios*
proportional: *analogos*
proportionally: *(kat') analogon*

propose: *protithesthai*
proposed, be: *prokeisthai*
proposition: *legomenon*
propound: *apodidonai*
prove: *deiknunai*
prove previously, in advance:
   prodeiknunai
prove to be: *anaphainesthai*
pure, be: *kathareuein*
put down: *tithenai*
puzzle: *aporia*
puzzle, be a: *aporein*
puzzled, be: *aporein*
puzzling: *aporos*

qualitatively: *(kata) poiotêta*
quality: *poion, poiotês*
quantify: *posoun*
quantitatively: *(kata) posotêta*
quantity: *poson*
quibble: *sophizesthai*

ratio: *logos*
rational: *logikos*
reading: *graphê*
reason: *aitia, aition*
reason falsely: *paralogizesthai*
reasoning: *epikheirêsis, logos*
recall: *hupomimnêskein*
reckon in addition: *proslogizesthai*
reduce (to impossibility): *apagein*
reduce: *kathairein*
reduction (to impossibility): *apagôgê*
reduction: *kathairesis*
redundantly: *(ek) parallêlou*
refer: *anagein, anapheirein,*
   apoteinesthai
refutation: *elenkhos*
refutation, in: *lutikos*
reject: *apogignôskein*
relatively: *skhetikôs*
relevant: *pros to*
remaining, be: *hupoleipesthai*
remark: *ephistanai*
remind: *hupomimnêskein*
reminder: *hupomnêsis*
resolution: *diakrisis*
resolve (thoroughly): *dialuein*
resolve: *luein*
responsible: *aitios*
rest: *êremia*
rest, be at: *êremein*
rest, bring to: *êremizein*
resting: *êremêsis*
result (vb.): *sumbainein*
result: *epiphora*
return (n.): *epanodos*

revolution: *kuklophoria*
revolving: *enkuklios, kuklophorêtikos*
ridiculous: *geloios*
right (also of angles): *orthos*
right as a whole, in its own: *kath'holon heauto*
right, in its own: *kath'hauto (kath'heauto)*
rotate: *peripheresthai*
rotation: *periphora*
route: *hodos*
rule (that): *axioun*
run together: *suntrekhein*

safe: *asphalês*
same time, at the: *hama*
section: *moira*
seeing: *horasis*
segment: *tmêma*
self-evidence: *enargeia*
self-evident: *enargês*
senses, in many: *pollakhôs*
separable: *khôristos*
separate: *khôrizein*
separate out: *aphairein*
separated, be: *diistasthai*
separateness: *khôrismos*
serious: *pragmateiôdês*
set forth: *katatattein*
set out: *horman*
set: *tithenai*
settle: *hidruein*
share: *koinoun*
shift: *meterkhesthai*
short of (something), be: *elleipein*
short, in: *(hôs) sunelonti (eipein)*
shortfall: *elleima*
show: *endeiknusthai, epideiknunai*
sign: *sêmeion*
signify: *dêloun*
similar: *homoios*
similar speed, at a: *homotakhôs*
similar speed, of a: *homotakhês*
similarity: *homoiotês*
similarly: *homoiôs*
simply: *haplôs*
simultaneously: *hama*
skip over: *huperbainein*
slow: *bradus*
slowness: *bradutês*
solution: *lusis*
solve: *luein*
sophistically: *sophistikôs*
sophistry: *sophismos*
sort (particular): *toionde*
soul, of the: *psukhikos*
sound (of an argument): *hugiês*

spatially: *topôi*
specifically: *idiôs*
specify: *aphorizesthai*
speculate: *theôrein*
speculation: *theôria*
speed: *takhos, takhutês*
speed, at the same (or a uniform): *isotakhôs*
speed, equal in, of equal (or uniform): *isotakhês*
speed, equal: *isotakheia*
spend (time): *poiein*
spherically: *sphairikôs*
spin round: *peridineisthai*
stand apart: *aphistanai, diistasthai*
stand still: *histasthai* (pf.)
start: *horman*
state (n.): *pathos*
statement: *legomenon, lexis, rhêton*
stipulate: *diatattesthai*
stop (intrans.): *pauesthai*
stop (n.): *stasis*
straight, straithway: *euthus*
strength: *bebaiotês*
strict sense, in a: *kuriôs*
strictly: *pantôs*
subject: *hupokeimenon*
subject to, be: *ekhein, hupopiptein*
substitute (A for B): *metalambanein (A ek B)*
substitute (B for A): *metalambanein (A eis B)*
subtract: *aphairein*
succeed (upon): *diadekhesthai*
sudden: *exaiphnês*
summary: *epidromê*
superimposed, be: *epharmozein*
support: *sunistanai*
supposed, be: *keisthai*
supposition: *keimenon*
surface: *epiphaneia*
surround: *periekhein*
switch: *metabainein*
syllogism: *sullogismos*
syllogize: *sullogizesthai*
sympathetic: *sumpathês*

take (X) along with (Y): *sumparalambanein*
take: *ekhein, lambanein, paralambanein*
take over: *epilambanein*
take up in turn: *metalambanein*
taken, able to be: *lêptos*
term: *onoma*
theorem: *theôrêma*
think: *noein*
think of: *epinoeisthai*

think up: *ennoein*
thinking: *noêsis*
this much: *tosonde*
thought: *gnômê*
touch: *haptesthai*
touch upon: *ephaptesthai*
tragic style, add in: *prostragôidein*
transfer: *metalêpsis*
transform: *alloioun*
transformation: *alloiôsis*
travel: *hodeuein*
traverse: *dierkhesthai*
traverse entirely: *diexerkhesthai*
treatise: *pragmateia*
true: *alêthês*
true together with, be: *sunalêtheuein*
true, be: *alêtheuein*
trust: *pisteuein*
truth: *alêtheia*
truth, lover of: *philalêthês*
turns, by: *(para) meros*
two kinds, of: *dittos*
type: *eidos*

unable: *adunatos*
undergo: *paskhein*
underlie: *hupokeisthai*
understandable: *eikos*
unevenly: *anômalôs*
unforced way, in an: *abiastôs*
ungenerated: *agenêtos*
uniformly: *homoiôs*
unit: *monas*
unsuitability: *anepitêdeiotês*

untainted: *akhrantos*
untraversable: *adiexitêtos*
useful: *khrôsimos*

vain, in: *matên*
vaporous: *pneumatikos*
variously: *polueidôs*
vault (of heaven): *hapsis*
vehicle (for the soul): *okhêma*
version: *diaskeuê*
vice versa: *anapalin*
volume: *onkos*
volume, of equal: *isoonkos*

walk: *badizein*
walking: *badisis*
water-creature: *enudron*
way: *tropos*
ways, in several: *pleonakhôs*
weakly: *malthakôs*
whole: *holos*
will: *boulêsis*
withdraw: *aphistasthai*
word: *lexis*
world: *kosmos*
world, in the: *enkosmios*
write a treatise: *pragmateuesthai*
write: *graphein*

yield: *sunagein*
yonder (side): *epekeinos*

zodiac: *zôidiakon*
zodiacal: *zôidiakos*

Subject Index

Achilles paradox, see Zeno
actual, of infinity, 948,23-24; see also
    infinity, potential
Alexander of Aphrodisias, on all-at-once
    change, 966,5-30; 968,6-969,9;
    978,35-979,7; on contact and coming
    to be, 997,30-998,3; on continuous
    magnitudes, 931,24-26; on the
    argument for divisibility from faster
    and slower things, 937,25-938,8; on
    the division of motion according to
    time and the parts of the moving
    thing, 974,25-29; on immediate
    change to a new state, 981,15-19; on
    the motion of partless things,
    1024,25-27; on non-consecutiveness
    of points, 930,11-931,7; on points and

lines, 929,10-18; on power, 941,22-
    942,24; on rotational motion,
    1022,16-22 (on the motion of great
    circles, 1026,5-11; on the rotation of
    the heavenly body, 941,22-942,18); on
    the soul's inseparability from the
    body, 964,14-966,14; on the paradox,
    'When did Dion die?', 983,25-984,2;
    on Zeno's fourth paradox (the
    stadium), 1017,18-21, 1019,27-32; on
    the reading of Aristotle's text at
    232a20, 936,21-28; on the reading of
    Aristotle's text at 233b6-7, 949,30-
    950,6; on the meaning of Aristotle's
    text at 235a18, 975,24-25
all-at-once change, 966,15-969,24;
    998,9-19; and burning (by the sun),

1011,29-1012,13); not at rest in any
first (or primary) time, 1009,21-
1011,8; time, motion, etc., as
properties of, 978,27-31; cf. also
changing thing

necessity, of two kinds, 942,5-8 (see also
power)
non-uniform motion, 999,25-1001,19;
and the divisibility of time, 937,25-
945,22
not-being, a sort of thing, 1030,5
now, having changed in, 982,3-984,2 (of
change between contradictories,
983,16-25; Alexander's corollary
concerning the paradox, 'When did
Dion die?', 983,25-984,2); division of
(Aspasius' view), 958,7-12; is
indivisible and one, the limit of the
past and future, 955,12-960,5
(indivisibility, 956,17-957,2;
960,8-20); and the motion of partless
things, 1027,13-31; no motion in,
960,23-962,22; nothing moves or is at
rest in, 960,23-962,22; moving thing
at a place equal to itself in, 1010,25-
1011,8; parts of (in the broad sense)
are now, 959,3-960,5; is partless,
926,19-21; and Plato, 982,6; exists
potentially in time, 1012,13-19; two
kinds (instant or temporal point [the
primary sense], and present time [the
broad sense]), 955,3-16 (now [in the
broad sense] contains the now in its
own right, 958,15-26)

one, two senses of (as indivisible, and as
single extended thing), 960,16-20
opinion, an incorporeal motion of the
soul, 965,2-21
oyster-like vehicle of the vegetative soul,
966,4-5

paradoxes concerning motion, change
between contradictories, 1020,9-
1022,4; a moving thing is in that from
which or that to which, 964,9-14; how
the parts of a moving thing move
(Eudemus), 973,26-974,10; a rotating
object and motion in the same place,
1022,7-1024,8 (Aspasius' view,
1022,14-15; Eudemus' view, 1022,15,
1024,5-8; Alexander's view, 1022,16-
22); 'When did Dion die?' (Alexander's
solution), 983,25-984,2; see also Zeno
part, no first part of changing thing,
987,11-29; two kinds of motion of,

1025,22-1026,17; moves incidentally
in a whole, 1025,9-12; of moving
thing as divisor of motion, 969,27-
974,13; of now (in the broad sense) is
now, 959,3-960,5; how parts of
moving thing move (Eudemus'
puzzle), 973,26-974,10; sum of
motions of parts equals motion of
whole, 970,26-971,9 (Eudemus'
version, 973,21-26)
partless things, definition of, 1025,14-15;
do not make a continuous magnitude,
925,22-926,21; move (or change) only
incidentally, 964,6-8; 1024,20-
1029,21 (Aspasius' view, 1024,27-
1025,2; Alexander's view, 1024,25-
27); cannot move or change (in their
own right), 962,25-964,8 (proof from
points and lines, 1028,3-31); could
move only if time were made of nows,
1027,13-31; cannot touch, 926,24-
927,9 (Eudemus' argument from
bisection, 930,35-931,6); the now is
partless, 926,19-21; the (numerical)
unit is partless, 926,19-21
past, see time, limit
perception, as evidence for change,
969,2-6
perfect (aspect of the verb), having
moved vs. moving, 933,13-934,30
Peripatetics, 923,3; 965,8.18
perishing, see coming to be
petitio principii, and the proof of the
indivisible now, 956,17-957,2
place, co-divided with other continua (as
that in respect to which change
occurs), 977,17-30; no first bit in local
change, 988,3-989,30; two senses of
(primary and broad), 955,9-12
Plato, on indivisible limit of time, 982,6;
allusion to the *Timaeus*, 946,26-27
pleasure, as the limit of activity, 928,17-
19
points, not consecutive, 927,26-928,27;
do not have extremes, 926,3-12; do
not make a line, 926,1-12
(Alexander's argument, 929,10-18;
Eudemus' argument, 928,28-929,9);
same kind of thing as lines, 928,15-
23; and the impossibility of motion of
partless things, 1028,3-31; are
quantities, 928,19-23
potential (as opposed to actual), of
infinite divisions, 1013,3-28; of
infinity, 947,28; 948,10-15; potential
infinity and Zeno's second paradox,
1015,21-26 (cf. 947,28); of nows in